Formal Aspects of Context

APPLIED LOGIC SERIES

VOLUME 20

SCOPE OF THE SERIES

Logic is applied in an increasingly wide variety of disciplines, from the traditional subjects of philosophy and mathematics to the more recent disciplines of cognitive science, computer science, artificial intelligence, and linguistics, leading to new vigor in this ancient subject. Kluwer, through its Applied Logic Series, seeks to provide a home for outstanding books and research monographs in applied logic, and in doing so demonstrates the underlying unity and applicability of logic.

The titles published in this series are listed at the end of this volume.

Formal Aspects of Context

edited by

PIERRE BONZON

Université de Lausanne, Switzerland

MARCOS CAVALCANTI

Universidade Federal do Rio de Janeiro

and

ROLF NOSSUM

Agder College, Kristiansand, Denmark

KLUWER ACADEMIC PUBLISHERS

DORDRECHT / BOSTON / LONDON

A C.I.P. Catalogue record for this book is available from the Library of Congress.

ISBN 978-90-481-5472-2

Published by Kluwer Academic Publishers,
P.O. Box 17, 3300 AA Dordrecht, The Netherlands.

Sold and distributed in North, Central and South America
by Kluwer Academic Publishers,
101 Philip Drive, Norwell, MA 02061, U.S.A.

In all other countries, sold and distributed
by Kluwer Academic Publishers,
P.O. Box 322, 3300 AH Dordrecht, The Netherlands.

Printed on acid-free paper

Printed in the Netherlands.

CONTENTS

EDITORIAL PREFACE

We welcome Volume 20, *Formal Aspects of Context.* Context has always been recognised as strongly relevant to models in language, philosophy, logic and artificial intelligence. In recent years theoretical advances in these areas and especially in logic have accelerated the study of context in the international community. An annual conference is held and many researchers have come to realise that many of the old puzzles should be reconsidered with proper attention to context.

The volume editors and contributors are from among the most active front-line researchers in the area and the contents shows how wide and vigorous this area is. There are strong scientific connections with earlier volumes in the series.

I am confident that the appearance of this book in our series will help secure the study of context as an important area of applied logic.

D. M. Gabbay

INTRODUCTION

This book is a result of the First International and Interdisciplinary Conference on Modelling and Using Context, which was organised in Rio de Janeiro in January 1997, and contains a selection of the papers presented there, refereed and revised through a process of anonymous peer review.

The treatment of contexts as bona-fide objects of logical formalisation has gained wide acceptance in recent years, following the seminal impetus by McCarthy in his Turing award address.

The field of natural language offers a particularly rich variety of examples and challenges to researchers aiming at the formal modelling of context, and several of the chapters in this volume deal with contextualisation in the setting of natural language. Others take a purely formal–logical point of departure, and seek to develop general models of even wider applicability.

There are twelve chapters, loosely connected in three groups. In the broadest of terms, the first four chapters are concerned with the formalisation of contextual information in natural language understanding and generation, the middle five deal with the application of context in various mechanised reasoning domains, and the final three present novel non-classical logics for contextual applications. Let us make some brief comments about each chapter.

In the first chapter, **Kees van Deemter** and **Jan Odijk** take a novel approach to context modelling in a text generation system, and point out a fertile common ground with current formal–logical theories of context. Next, **Harry Bunt** goes on to investigate communicative activity in dialogues, and analyses the aspects of context addressed therein. In conclusion, he calls for a unified representational formalism encompassing four specific dimensions of context. Reporting an empirical study, **Julia Lavid** makes considerable progress in exploring the relationship between specific contextual variables and the phenomenon of thematisation in discourse, overcoming some limitations of previous studies. **Mark Galliker** and **Daniel Weimer** investigate a corpus of verbally discriminatory texts in search of explicit indicators for implicit meaning, finding by co-occurrence analysis that sentences with person categories of devaluative connotation only on the basis of their context, exhibit a high frequency of particles and modal words.

In the fifth chapter, **Alessandro Cimatti** and **Luciano Serafini** apply belief contexts to mechanise multicontextual reasoning in an interactive inference system, obtaining automated solutions to several versions of the Three Wise Men puzzle. Next, **Paul Piwek** and **Emiel Krahmer** reformulate van der Sandt's theory of presuppositions as anaphors in terms of constructive type theory, and obtain an attractive mechanism for resolving presuppositions against world knowledge, in particular relating to the bridging phenomenon of Clark. **Vagan Terziyan** and **Seppo Puuronen** propose an interpretation of objects in a semantic network as contexts, and present a general framework for solving several kinds of problems relating to contextual knowledge. In pursuit of generality in machine learning, **Pierre Bonzon** develops a contextual learning model where discovery of re-usable object-level concepts is feasible. His approach profits from the general nature of lifting theories for contexts. **Leendert van der Torre** and **Yao-Hua Tan** devise a contextual deontic logic, in which the notorious contrary-to-duty paradoxes of other deontic logics are overcome by distinguishing between violations an exceptions, which are treated as factual and overridden defeasibility, respectively.

In the tenth chapter, **Fausto Giunchiglia** and **Chiara Ghidini** build on the hierarchical multilanguage belief systems of Giunchiglia and Serafini and construct a local model semantics for certain epistemic systems of logic. Continuing along the same line, **Luciano Serafini** and **Chiara Ghidini** proceed to develop a model theoretic semantics for federated databases, using an extension of the semantics developed in the previous chapter. In the final chapter, **Dov Gabbay** and **Rolf Nossum** present a general context theory, give a fibred semantics framework for it, and show how some context formalisms in the literature fit in as special cases.

The editors wish to thank the anonymous referees for their efforts, and the competent staff at Kluwer for making the appearance of this book possible. In particular, we wish to thank **Jane Spurr**, without whose skill and persistence this book project could never have been completed.

P. Bonzon, M. Cavalcanti and R. Nossum

KEES VAN DEEMTER AND JAN ODIJK

FORMAL AND COMPUTATIONAL MODELS OF CONTEXT FOR NATURAL LANGUAGE GENERATION

1 INTRODUCTION

Context-dependent interpretation has taken centerstage in the theatre of language interpretation. The interpretation of personal pronouns, for example, is known to depend on the linguistic as well as the nonlinguistic environment in which they appear. Moreover, it has become clear that very similar kinds of dependence on context apply to many other phenomena including, among other things, the contextually restricted interpretation of a full Noun Phrase, the determination of the 'comparison set' relevant for the interpretation of a semantically vague predicate, the determination of so-called 'implicit arguments' of words like *local* and *contemporary*. (For references to the literature, see [van Deemter and Odijk, 1997].) Inspired by this growing body of work, dependence on linguistic context has become the cornerstone of the so-called dynamic theories of meaning (e.g. [Kamp and Reyle, 1994]). These theories characterize the meaning of a sentence as its potential to change one 'information state' into another, and it is this dynamic perspective on which current natural-language interpreting systems are beginning to be based.

The importance of context for language interpretation raises the question to what extent context modeling is relevant for natural-language generation. In some sense, the relevance of contextual information for generation follows from the relevance of this information for interpretation, since the discourse generated has to be interpreted by the user of the system. For example, when a pronoun, a demonstrative ('this composition', 'those fugues'), or a definite description ('the man') is generated, its linguistic context determines how it will be interpreted by the user. But there is more. For example, context determines what may be called informally the *relevance* of a given item of information at a given point in the discourse. (Not all true statements on the topic of this paper could felicitously be expressed here, for example.) Furthermore — and this is where the relevance of context for speech comes in — those parts of a sentence that are responsible for novel information are likely to be accented in speech. Thus, both content and form of *any* utterance are affected by linguistic context.

Natural language generation systems are beginning to take these phenomena into account. Especially the generation of nominal expressions has attracted much attention (see e.g. [Dale and Reiter, 1995] for an overview)

P. Bonzon, M. Cavalcanti and R. Nossum (eds.), Formal Aspects of Context, 1–21.

and, to a lesser extent, the connection between (de)accenting and information status (e.g. [Prevost and Steedman, 1994]). A principled perspective on all the different roles of context in natural language generation has not been advanced, however. The present paper tries to chart the problem and compare some possible approaches.

The first five sections of this paper will explain informally how linguistic context can be taken into account by a system that generates discourse in an incremental (as opposed to a plan-based) fashion. We will use the so-called Dial-Your-Disc (DYD) system as an example [Collier and Landsbergen, 1995; van Deemter and Odijk, 1997] because, on account of the 'incremental' approach to language generation of this system (outlined in Section 4), context modeling plays an even more central role in it than in most other natural language generation systems. The main novelty of this paper has to wait until section 6, which contains a comparison between the computationally motivated Context Model of the DYD system, on the one hand, and a number of context models that have come up in formal semantics and artificial intelligence, on the other. In this section, it will be argued that the framework of the *Ist*-theory of context (see e.g. [McCarthy, 1993]) could facilitate some types of reasoning that can be used by a dialogue system that exploits 'domains of discourse' in its generation and understanding of contextually appropriate quantified expressions.

2 A SKETCH OF THE DYD SYSTEM

The DYD system produces spoken monologues derived from information stored in a database about W. A. Mozart's instrumental compositions. The purpose of the monologue generator is to generate from these data a large variety of coherent texts, including all information required for a correct pronunciation. A generator like this could be part of an electronic shopping system, where users can express their interest in a certain area without being completely specific, and where the system provides information and 'sales talk'. The way in which users can indicate their areas of interest will not be discussed in this paper, which focuses on language and speech generation. A simplified database representation of a recording could be:

KV 309
DATE 10/1777 - 11/1777
SORT piano sonata
NUMBER 7
PERFORMER Mitsuko Uchida
PLACE London
VOLUME 17
CD 2
TRACK 4

Our current system could, after a client has shown interest in the composition described by the above database object, come up with the following text. (See [van Deemter and Odijk, 1997] for more examples.)

> *You are now going to listen to a fragment of piano sonata number seven, K. 309, played by Mitsuko Uchida. It was composed for Rosine, the daughter of the court musician and composer Christian Cannabich in Mannheim. The composition was written from October 1777 to November 1777. The recording of K. 309 was made in London. It starts at track four of the second CD of volume 17.*

3 SYSTEM ARCHITECTURE

An important system requirement is that a large variety of texts can be produced from the same database structures. Presentations are generated on the basis of database information by making use of *syntactic templates*: structured sentences with variables, i.e. open slots for which expressions can be substituted. These syntactically structured templates indicate how the information provided by a (part of a) database object can be expressed in natural language. The required variety is achieved by having many different templates for the same information and by having a flexible mechanism for combining the generated sentences into texts. In addition, information is available that does not fit in the uniform database format. This is canned text, represented by object-specific templates expressing this information. In the example text in section 2, the mention of Rosine is canned text. The remainder of the presentation has been generated by general templates. A template can be used, in principle, if there is enough information in the database to fill its slots. However, there are extra conditions to guard the well-formedness and effectiveness of presentations. For example, certain points in the discourse are more appropriate for the expression of a certain bit of information. Thus, it is important for the system to maintain a record showing which information has been expressed and when it has been expressed. This record, which is called the *Knowledge State*, will be part of DYD's *Context Model*. (For an overview of the different components of DYD's Context Model, see Section 6.1.)

Many variations of the above presentation are possible. The system can, for instance, start with mentioning the date of composition, or information could be added that contrasts this composition with a previous one. Also, there are various ways to refer to the composition being discussed, for instance by name (*K. 309*), with a definite noun phrase (*the composition*), or with a pronoun (*it*). The appropriateness of a referring expression depends, among others, on the existence and kinds of references to the referred object in previous sentences. This means that it is important to maintain a record

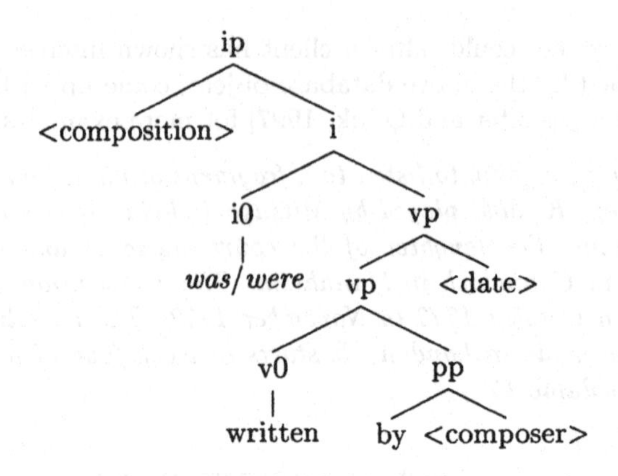

Figure 1. Template for a sentence like "K. 32 was written by Mozart in March 1772". *Composition, composer, date* are the variables of the template. *Was/were* indicates that a choice must be made, depending on the subject. The labels *vp, v0* and *pp* stand for *verb phrase, verb* and *prepositional phrase*, resp. A sentence (*ip*) is headed by an abstract *inflection* node (*i0*), which can contain auxiliary verbs and shows the normal projections in accordance with X-bar Theory.

of which objects have been introduced in the text, and how and when they have been referred to. This record will be called the *Discourse Model*, which is also a part of the Context Model.

As was mentioned above, templates in our system are structured sentences with slots. A simplified example is given in Figure 1. The slots *composition, composer, date* are to be filled with structured expressions that contain database information. This is done with other, smaller, templates. One reason for having templates of syntax trees is that the prosodic module of our system, in which accents and pauses are determined, needs the syntactic structure of sentences. Another is that this structure enables one to write rules to check the syntactic correctness of the sentences after slot-filling and to adjust them if necessary.

The architecture of our system is shown in Figure 2. It displays three modules: *Generation*, *Prosody* and *Speech*. The module *Generation* generates syntax trees on the basis of the database, a collection of templates, and it maintains the Context Model. The module *Prosody* transforms a syntax tree into a sequence of annotated words, the annotations specifying accents and pauses. The module *Speech* transforms a sequence of annotated words into a speech signal.

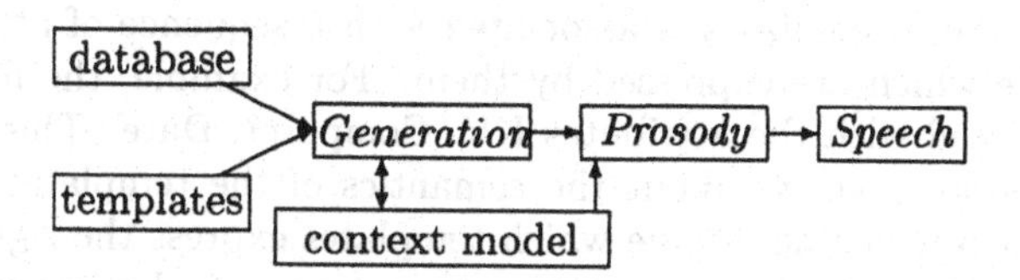

Figure 2. System Architecture: *Generation* generates syntax trees on the basis of the database and a collection of templates, and it maintains the Context Model. *Prosody* operates on syntax trees and the Context Model to produce a sequence of annotated words. *Speech* takes the output of *Prosody* and outputs an acoustic realization.

4 TEXT GENERATION

It will be discussed here how texts can be generated, concentrating on three aspects: (1) the generation of a text; (2) how to achieve its coherency, and (3) how to achieve the required variation. We will assume that information about K. 32 is stored as follows:

COMPOSER W.A.Mozart
KV 32
TITLE Galimathias Musicum
DATE 3/1766
SORT quodlibet

4.1 Sentence generation

As explained in the previous section, sentences are generated by means of syntactically structured templates. A template indicates how the meaning of a database record can be verbalized. Since there are various ways to verbalize the content of a record, and many ways to group information from different records into one verbalization, this will lead to a large number of possible sentences for conveying the database information.

In the examples below we will use the template introduced in figure 1 (repeated in (1a)) and a new one (1b). For expository convenience, we will represent only the terminals of templates. Variable parts (e.g. <composition>) will be represented between angled brackets. An example is

(1) a <composition> was/were written by
<composer> <date>
b We will now present information about
<composition>.

Each of these templates is associated with a sequence of attributes in the database which are expressed by them. For example, the first of the two is associated with the attributes KV, Composer, Date. This association is all the system knows about the semantics of the templates and it is this association which determine which templates express the right information to be usable in a given situation. A sentence can be constructed from a template by filling the variable parts. Example sentences derived from the two templates could be:

(2) a We will now present information about 'Galimathias Musicum'.
b K. 32 was written by Mozart in March 1766.

The fact that templates are structured objects makes it possible to formulate various conditions on the form of variable parts. In this way, it is possible to avoid the generation of incorrect sentences such as:

(3) a *We will now present information about he.
b *It were written by him when Mozart was only ten years old.

In the first example the pronoun *he* is selected to express the composition, but that is wrong in two respects. First, *he* is not an appropriate pronoun for compositions (but only for persons), and second, in the sentence given its form should be *him*, not *he*. In the second sentence, the choice for the finite verb *were* is incompatible with the singular subject *it*, and the co-occurrence of *him* and *Mozart* suggests that these expressions refer to two different persons, though they actually refer to one and the same person.

Since templates are structured objects, conditions guaranteeing the appropriate choice of pronouns can refer to information contained in these structures (e.g. that *he* refers to persons and that *he* is governed by the preposition *about*).

Similarly, it can be read off the syntactic structure that the pronoun *it* is the singular subject of the second sentence and that therefore the finite verb should be *was*. The infelicitous choice of *him* and *Mozart* is prevented by a more complex condition on the proper sentence-internal distribution of pronouns, proper names and other expressions. This condition is a version of the so-called *Binding Theory* (see Chomsky [1981; 1986]), which is crucially formulated in terms of configurations in syntactic structure.

4.2 Discourse coherence

Now we are able to use a large variety of sentences to convey the relevant information, but it is as yet unclear which sentences should be used in a given situation. This problem is solved in two steps. First of all, it has to be determined what is going to be said. This is determined during the dialogue, where the user can indicate a preference for less or more elaborate monologues, that is, monologues that express a smaller or a larger set of attributes and relations from the database. This preference is stored in the *Dialogue State*, a part of the Context Model in which all those properties of the dialogue history are recorded that are relevant for monologue generation.

Secondly, a selection has to be made from all templates in such a way that the text generated conveys all and only the required information. A minimal requirement is that those templates are selected which are able to convey the relevant information. An additional requirement is that, under normal circumstances, the same information is presented not more than once. Furthermore, the form in which this information is presented should vary to avoid stylistic infelicities. These requirements have been incorporated in the text generator, which also has as its task to present the sentences in such a way that the text shows a certain coherence. Information should be grouped into convenient clusters and presented in a natural order. Clustering is achieved by means of the so-called *Topic State*. For each paragraph of the monologue, the Topic State, which is another part of the Context Model, keeps track of the topic of the paragraph, which is defined as a set of attributes from the (music) database. For example, a paragraph may have 'place and date of performance' as its topic and then only those templates can be used that are associated with the attributes 'date' and 'place'.

We will not deal here with the exact mechanisms of the text generator, but the idea is the following. Each template 'attempts' to get a sentence generated from it into the text. Whether this succeeds depends on the information conveyed by the sentence, which information has been conveyed earlier, and whether the sentence can find a place in a natural grouping of sentences in paragraphs. The method is characterized by the fact that only local conditions on the Context Model and the properties of the current template determine whether a sentence is appropriate at a certain point in the text. No global properties of the text are considered and no explicit planning is involved. For a more detailed description of this approach we refer to [Odijk 1994].

Since the 'gaps' in the templates used by the generator are, in the vast majority of cases, noun phrases, finding the right noun phrase to denote a particular individual or set of individuals is one of the most important challenges for the generation of coherent discourse. This includes the generation of contextually appropriate proper names, definite descriptions (i.e. noun phrases introduced by the definite article *the*), pronouns (e.g. *he, it, his, they, himself*), 'demonstrative descriptions', i.e. noun phrases introduced by the determiners *this, that, etc.* (e.g. *this quodlibet*), various 'relational descriptions' (e.g. *his sister*). Here again, there are strict rules which determine when the use of such a device is appropriate. If no such rules are incorporated in the text generator, it is possible to generate deviant texts such as:

(4) a We will now present information about it. K. 32 was written by him when the composer was only ten years old.
b We will now present information about the quodlibet. 'Galimathias Musicum' was written by Mozart when he was only ten years old.

In the first text, the pronouns *it* and *him* are used without a proper antecedent, and in both sentences the definite descriptions *the composer* and *the quodlibet* are also used incorrectly. Thus, it is necessary to formulate rules which guarantee the proper usage of such anaphoric devices.

For each type of expression conditions must be formulated which determine their proper use in a text. Apart from pronouns and definite descriptions as discussed above, indefinite expressions and various quantified expressions must be dealt with as well. Plural noun phrases and negation introduce yet other complexities which must be dealt with adequately.

In addition, there are various conditions on the 'distance' between an antecedent and an anaphoric device. The determination of the exact formulation of 'distance' is a complex issue (a definition in terms of the number of preceding sentences is in general too simplistic), but such conditions must be incorporated to achieve the appropriate coherence in a text (see [Grosz *et al.*, 1986; Dorrepaal, 1990]).

As we have seen in Section 3, an important part of the Context Model (see Figure 1) is a *Discourse Model.* Starting with an empty Discourse Model, each candidate sentence adds discourse referents and relevant associated information to this model. For example, the Discourse Model may record that a certain description (e.g. *this composition*) has occurred as the xth and $x + 1$st word of the yth sentence of paragraph number z of the uth monologue that has occurred during a given user-system interaction. Rules for anaphora establish the antecedents for anaphora, and afterwards it is checked whether the resulting Discourse Model is well-formed (e.g. by checking whether each pronoun has an antecedent, whether definite descriptions have been used appropriately, etc.). If the Discourse Model is found to be well-formed, the candidate sentence can be used as an actual sentence. If not, a different candidate sentence is subjected to examination, etc. We will see that very similar rules, which are also based on the information in the Discourse Model, are used to determine whch words in the sentence are to be accented.

5 PROSODY AND SPEECH

Generating acceptable speech requires syntactic and semantic information that is hard to extract from unannotated text. this makes 'text-to-speech' syntesis a difficult problem. In the present, 'concept-to-speech', setting, however, speech generation is helped by the availability of syntactic and semantic information. When the generation module outputs a sentence, the generated structure contains all the *syntactic* information that was present in the template from which it results. Moreover, the Discourse Model, as we have seen, contains *semantic* information about the sentence. In what follows, we will show how both kinds of information are used to solve one of most important problems in the generation of intonationally adequate speech, namely the proper location of pitch accents. The (prosodic) question of accent location is discussed in Section 5.1, while the (phonetic) question of how accents are realized is discused in Section 5.2.

5.1 Prosody

What words in a text are to be accented? At least since Halliday's [1967], students of Germanic languages have known that one factor that must be taken into account to answer this question is *information status*: in these languages, 'new' information must be stressed. This idea was later corroborated by experimental research (e.g. [Terken and Nooteboom, 1987]). Existing speech synthesis systems (e.g. Bell Labs' Newspeak program) have capitalized on this insight, by de-stressing all content words that had occurred in the recent past. Yet, these systems are generally judged as still stressing too many words [Hirschberg, 1990]. Our own approach improves upon earlier approaches to accent placement by combining syntactic and semantic information. In particular, we have redefined the contextual notions of givenness and newness in such a way that these notions are properties not of individual words, but of entire phrases. (See [van Deemter, 1998b] for a recent account.) These definitions are then combined with a version of Focus-Accent theory to determine the exact word at which the accent must land. We will first discuss the semantic part of the problem, which may be rephrased as the question of which slot fillers are 'in focus', and then the syntactic part of the problem which deals with the question of what word in a focussed phrase must be accented.

Givenness and newness redefined

Givenness and newness are customarily defined in ways that do not do full justice to the semantic nature of these distinctions: In most approaches, a word is considered given if it is either identical to a word that has already occurred, or a slight morphological variation of such a word. However, inspection of the relevant facts suggests strongly that words of very different forms may cause a word to have 'given' status. For example, the word *wrote* can not only become 'given' due to an occurrence of *wrote*, *write*, etc., but also due to an occurrence of the word *compose*. In addition, givenness is not restricted to individual words. For example, an occurrence of *K.32* or of *this composition* may become 'given', and hence de-stressed (de-accented) due to an earlier reference to K. 32, as when 5a is followed by 5b.

(5) a You have selected K.32.
b You will now hear K.32\this composition.

Thus, semantic theories are natural suppliers of proper definitions of newness and givenness. Moreover, de-stressing and pronominalization occur in roughly the same environments, namely those in which an expression contains 'given' information. This suggests that both may be viewed as reduction phenomena that are caused by semantic redundancy. For these and similar reasons it has recently been proposed that theories of anaphora should be used in accent prediction algorithms. The Discourse Model presents itself as a natural candidate to implement this idea, since it contains all the relevant information. In particular, it says, for each referentially used Noun Phrase, whether and where in the discourse the object that it refers to was described earlier. If such an 'antecedent' for an expression is found earlier in the same paragraph, the expression is considered 'given' information (i.e. it is not 'in focus'). If not, it is considered 'new' (i.e. it is 'in focus').

A version of Focus-Accent theory

Focus-Accent theory was first conceived by Ladd and others [Ladd, 1980], and later refined by various authors. Our own implementation of Focus-Accent theory, which is sketched below, extends an implementation by Dirksen [Dirksen, 1992] by adding semantic considerations to Dirksen's syntactic account. Interested readers are referred to [Dirksen, 1992] and [van Deemter, 1998b] for specifics.

The basic insight of Focus-Accent is the idea that the syntactic structure of a sentence co-determines its 'metrical' structure. Metrical structure is most conveniently represented by binary trees, in which one daughter of each node is marked as *strong* and the other as *weak*. Metrical structure determines which leaves of the tree are most suitable to carry an accent on syntactic grounds. Roughly, these are the leaves that can be reached through a path that starts from an expression that is 'in focus', and that does not contain *weak* nodes. More exactly, *if* a given major phrase is 'in focus', it is also marked as *accented*, and so is each strong node that is the daughter of a node that is marked as *accented*. Accent is realized on those leaves that are marked as *accented*. However, there may be several obstacles

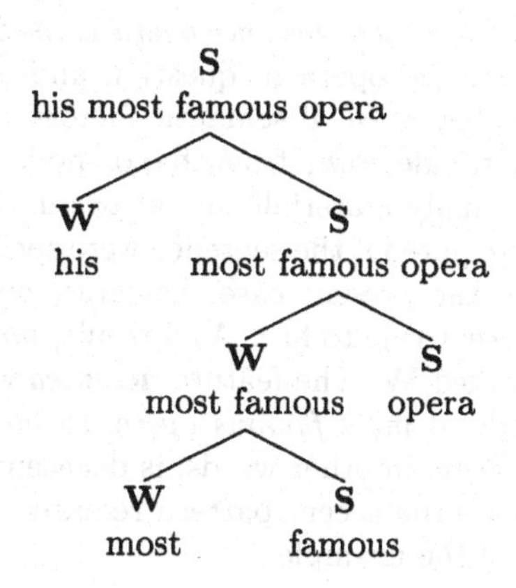

Figure 3. Example of a metrical tree

that prevent this from happening. Leaves may end up unaccented in several circumstances. For example,

(a) A major phrase is marked $-A$ if it is not in focus.

(b) A leaf x is marked $-A$ if there is a recent occurrence of an expression y which is semantically subsumed by x.

(c) A leaf is marked $-A$ if it is lexically marked as unfit to carry an accent that is due to informational status. (Examples: *the*, *a*, some prepositions.)

Only the first of these cases will be exemplied below, by means of a simple example. The result of an $-A$ marking is that the so-called *Default Accent rule* is triggered, which transforms one metrical tree into another:

> **Default Accent rule:** If a *strong* node n_1 is marked $-A$, while its *weak* sister n_2 is not, then the *strong/weak* labeling of the sisters is reversed: n_1 is now marked *weak*, and n_2 is marked *strong*.

Consider the discourse *Mozart wrote many operas. Die Zauberflöte is his most famous opera.* Note that the word *opera* in the second sentence occurs 'in a context' where the concept 'opera' has just been mentioned. This is a trivial application of clause (b), (*opera* semantically subsumes *operas*, and the previous sentence counts, by definition, as recent) which will mark the *opera* node as $-A$. The effect of this marking can be explained by means of the metrical tree figure 3. In English, it is usually the right daughter of a mother node that is *strong*, which explains the **W/S** marking in the tree. The Noun Phrase ***his most famous opera*** is used ('predicatively') to make a statement about the opera in question and is therefore in focus, and therefore labeled as *accented*. If semantic factors did not intervene, the feature *accented* would 'trickle down' from the top node, always following the **S** daughter of a node, finally materializing on *opera*. This is the accenting pattern that would be required if the sentence were spoken without previous mention of operas. In the present case, however, *opera* is marked $-A$, causing the Default Accent rule to fire. As a result, ***most famous*** is marked **S** and *opera* is now marked **W**. The feature *accented* will, therefore, trickle down from the top node to ***most famous opera***, to ***most famous***, and then to *famous*. The word *opera*, in other words, is deaccented and the accent is shifted to *famous*. This is the accent pattern required when the sentence is spoken in the context of the example.

5.2 Speech

The *Prosody* module of the system determines for every word in the monologue whether or not it occurs accented (*notation:* '+' or no marking), and it determines for every word boundary whether it is accompanied by a major prosodic boundary, a minor prosodic boundary, or by no boundary at all (*notation:* 'p2' or 'p1' or no marking). For example, the prosodic module may output an enriched sentence such as the following:

(6) This sonata for +violin and +piano *p*1
was written in +Salzburg *p*2 in +1735.

This prosodically enriched sentence is then passed on to the speech module, whose job it is to 'realize' this abstract structure in sound.

Ideally, the speech module consists of two independent parts, one of which takes care of segmental information and the other realizes accenting and phrasing. For present purposes, let us assume that all the segmental information is in place[1] and focus on the suprasegmental information, and especially on accenting. Accents are realized in accordance with the IPO model of Dutch intonation ['t Hart *et al.*, 1990]. This model has been applied to English [Willems *et al.*, 1988], and this work was used to extract rules for synthesis, which are applied in roughly the following way.

First, the prosodically enriched sentence is transformed into a structure in which the abstract information concerning accenting is 'interpreted' in terms of Rises and Falls. For instance, a sequence of accents is typically represented by a series of so-called 'pointed hats', followed by one 'flat hat', where the 'flat hat' represents the last two accents in the sequence. A 'pointed hat' designates an abrupt Rise in (F_0) pitch, immediately followed

[1] For the future, a segmental module is envisaged that makes use of diphones, i.e. recorded transitions between phonemes. The current system, however, makes use of the commercially available DECTALK system, which is based on synthesized formants.

by an abrupt Fall, whereas a 'flat hat' has an intermediate phase in which pitch remains equal.

Later modules transform this representation into one that is even closer to the physical level, by taking into account that accents are superimposed on a background of pitch declination. For example, a pitch accent that occurs at the beginning of a sentence typically has a higher 'peak' than one that occurs close to the end of the same sentence. Finally, a particular 'speaker' has to be chosen, with its own characteristic pitch range, timbre, etc. It is only at this stage that all the properties of the speech sound have been determined and that the generation of the monologue is finished.

6 CONTEXT MODELLING

6.1 Context modelling in DYD

The preceding sections have provided a highly informal overview of the way in which spoken monologues are generated in the DYD system and of the various aspects of linguistic context that play a role in the generation process. (For more detail and explanation, see [van Deemter and Odijk, 1997].) An overview of the different parts of DYD's Context Model, including some that have not been mentioned so far, may be helpful. DYD has used a Context Model that has four submodels: the Knowledge State, the Topic State, the Context State, and the Dialogue State. The *Knowledge* State keeps track of what information has been conveyed, and when. The *Topic* State keeps track of which topics have already been dealt with. The *Context State* (one part of the Context Model) keeps track of (i) the objects introduced in the text and their location, and (ii) linguistic expressions occurring in the preceding text.[2] Finally, the *Dialogue* State keeps track of which recordings were selected before, what the current recording is, and what kind of monologue should be generated (i.e. less or more elaborate).

For ease of reference we have put these submodels in Table 1. The table makes an effort to cluster elements of the Context Model that belong together, but a certain arbitrariness is unavoidable. We have seen in the Introduction of this paper that context modelling is essential for any system that performs natural language generation. Note, however, that DYD, as a result of its *incremental* approach to language generation, stresses some aspects of context modelling that are of much less importance in plan-based approaches to generation. These aspects, which have to do with the information that happens to be presented earlier in the monologue and the database attributes associated with this information, is modeled through

[2] The latter includes, for example, a record of the set of concepts that have been used in the current + previous sentence, which is needed for the deaccenting algorithm of Section 5.1.

Table 1. Parts of the Context Model and the information they contain.

Submodel	Contains
Knowledge State	What info has been presented where, and how explicitly
Topic State	Administration of current and previous topics
Context State	Discourse Model, templates/words used, recently used concepts, etc.
Dialogue State	recording (part) selected; history of previous selections, required degree of elaborateness

the Knowledge State and the Topic State. It could be argued that this incremental approach makes DYD an especially good model of *spontaneous* speech, where relatively little of what is said has been planned in advance.

6.2 Context Modelling in AI

A linguistic context, viewed as the situation that obtains at a given point in a discourse (e.g. a monologue), is a context like any other. Consequently, one might use a general theory of context to shed light on it. In particular, one might try to use the formalism of McCarthy/Guha/Buvac (sometimes called '*Ist*') theory to express truths about linguistic contexts (e.g. [McCarthy, 1993]) in such a way that all the aspects of the context that are relevant for generation come out. The key form of expressions in the Ist-theory is that of $Ist(c, \varphi)$, which can be read as saying that φ is true with respect to c. Now let c be the context that obtains after the sentence *Mozart composed K.280 in 1780* has been generated. Since, in the Ist-theory, contexts are first-class citizens, we can now say various things about c, and then use the Ist-theory to say that a second sentence (for instance, *It is a sonata*) is expressed in c. The notation DE(c) stands for the set of 'discourse entities' (i.e. objects or sets of objects introduced in the text) associated with c:

Text(c) = *Mozart composed K.280 in 1780,*[3]
DE(c) = {W.A.Mozart, K.280},

The 'Text' function might contain the entire preceding discourse or, more specifically, only the words contained in the previous sentence (unless it occurred in the previous paragraph), since these are the ones that are actually taken into account in DYD's implementation of the deaccenting rule (b) (Section 5.1). The 'DE' predicate plays the role of (the main part of) DYD's so-called Discourse Model, noting which objects in the database have been referred to in the monologue. This information can be exploited when the second utterance, *It is a sonata*, is interpreted 'in the context of' c:

Ist(c, *It is a sonata*)

Other properties of the context, of the kind that is relevant for language and speech generation, as we have seen, can be treated along the same lines:

Info-presented(c) = {Composition-Date}
Topic(c) = {Place and date of performance}
Dial(c) = { First movement of K.280, $\emptyset$, elaborate }

'Info-presented' and 'Topic' represent the information contained in DYD's Knowledge State and Topic State respectively. 'Dial' represents the part of the recording under discussion, the set of recording parts presented earlier in the dialogue (in the present example this is the empty set), and the fact that the user has asked for an elaborate, rather than a brief presentation. In principle, much more information about c could be encoded. For example, to generate personal pronouns, one might express that Speaker(c) = DYD, etc. Here, however, we will limit ourselves to the information represented in DYD's context model.

This example suggests that DYD's Context Model may be mirrored in the Ist-theory. An important aspect of 'linguistic' contexts that is not a standard element of the Ist-theory[4], however, is context-change. Linguistic contexts change during processing: discourse entities are added, objects and expressions move into and out of focus as more and more text is generated or interpreted. Context-change is the *raison d'être* for so-called dynamic theories of meaning, of which Discourse Representation Theory (DRT) is one of the best-known examples. Taking this into account, one might envisage an extension of the Ist-theory in which DRSs are contexts, and where DRT's context-change mechanism is exactly mirrored in the Ist-theory. For

[3]This might also be written as Ist(c_0,*Mozart composed K.280*), where c_0 is the context that obtains at the beginning of the monologue.

[4]See the treatment of question-answering contexts in [Buvač, 1996] for a first attempt at incorporating context-change into the Ist-theory.

example, one would need an 'update' operator '+' to say how a context c is changed when the (nonambiguous) sentence S has been processed in c:

$$c + S = c'$$

and one might define the notion of a 'continuation' using a recursion based on this operation:

c_x is a continuation of c_y iff
$\exists S : c_y + S = c_x$ or
$\exists S \exists c_z : c_y + S = c_z$ and c_x is a continuation of c_z.

Making use of these notions, one could express truths of the following kind:

If c' is a continuation of c
then $\mathrm{DE}(c) \subseteq \mathrm{DE}(c')$.

This formalizes the idealization - which is common in most linguistic theories of context - that the linguistic context 'grows' as we move from left to right through a discourse. Using such extensions, it might be possible to encode all of Discourse Representation Theory using the formalism of the Ist-theory. This may be a useful exercise, which can lead to a better understanding of the peculiarities of *linguistic* context, *vis-á-vis* other kinds of contexts. But it also raises the question of whether we might have used DRT as a backbone for DYD's Context Model. This question will be answered in the negative in the next section, which focuses on DRT.

6.3 Context Modelling in DRT

It might be thought that DRT could have provided us with the kind of context models we need. It is true that DRT's so-called Discourse Representation Structures (DRSs) were specifically designed to represent the contextual information that is relevant for the interpretation or generation of anaphoric (e.g. pronominal) material in a discourse [Kamp and Reyle, 1994]. In the setting of DYD, DRT could take the form of a context model containing a series of sub-DRSs, the first of which contains information extracted from the dialogue that has led up to the selection of the first composition plus the monologue following it, and so on. Thus, the context model that has been built up at the start of the $n + 1$-st monologue could be represented as follows:

```
CONTEXT MODEL
   DRS_OF_DIALOGUE_1
   DRS_OF_MONOLOGUE_1
    .....
```

```
DRS_OF_DIALOGUE_n
DRS_OF_MONOLOGUE_n

DRS_OF_DIALOGUE_n+1
```

where DRS-OF-DIALOGUE$_i$ (DRS-OF-MONOLOGUE$_i$) contains contextual information extracted from the ith dialogue (monologue). However, setting up structures of this kind would have required all kinds of information that are neither routinely represented in existing versions of DRT, nor trivial to calculate on the basis of them. For example, DRSs do not normally contain a representation of their subject matter (their 'topic') and it would not be a trivial matter to deduce this information from the truth conditions of the DRS.[5] The same is true for the degree of explicitness with which information in a DRS is presented. Furthermore, standard versions of DRT do not contain information about the exact place of occurrence of expressions (as does DYD's discourse structure), nor do they contain information about paragraph structure. Of course, information of all these kinds might be added. The result would be a new, extended version of DRT, which would complicate drastically the formal basis of this theory.[6] Moreover, conventional DRSs (e.g. [Kamp and Reyle, 1994]) contain plenty of semantic information that is not immediately relevant for current (i.e. generative) purposes. In particular, they do not have to encode the truth conditions of the sentences involved, since these are encoded in the mapping between templates and attibutes of the database (Section 4.1). In other words, DRSs contain both less and more than what is needed for language generation.

6.4 Potential advantages of using the Ist-theory

What has been said in the previous sections will have clarified why the DYD system has not implemented a pre-existing theory of contextual information. As indicated in section 6.2, however, it might have been possible to make use of an *extended version* of the Ist-theory, especially if it included a way of representing context-change. But what could one gain by making use of (an extended version of) the Ist-theory? In what follows we will try to answer this question. In particular, we will argue that a dialogue system might profit from being able to reason *about* contexts, and that this is an important potential advantage that the Ist-theory could have when applied to natural language processing.

[5] In particular, the set of topic of a discourse does not always equal the union of the sets of topics of its parts [Demolombe and Jones, 1995]. For example, one part of the discourse may be about friendship, another about injustice, and the combination may cause the discourse as a whole to be about betrayal.

[6] In particular,these extensions would cause the meaning of DRSs to be something much more complex than a relation between assignments.

Imagine the following fragment of a dialogue, as it may occur between people who have met at a conference.

> **A:** Everyone knows Montague Grammar.
> **B:** You're wrong. Plenty of people have never heard of Montague Grammar.
> **A:** I meant to say that everyone *at this conference* knows Montague Grammar.

Let us assume that Noun Phrases such as *everyone* can quantify[7] over the elements in a set that is smaller than that of all individuals in the domain. Thus, a sentence such as *Everyone knows Montague Grammar* will tend to quantify over the individuals in some contextually salient domain of discourse. A domain of discourse is a specific kind of context, and one that is very important in natural language. It can be determined by all sorts of factors, including the linguistic context (e.g. the set of objects introduced earlier in the dialogue) and the nonlinguistic context. In the case of our example, possible candidates for the role of domain of discourse might be, for example,

> (1) The set of all educated people,
> (2) The set of all people who are working in the topic area of the conference,
> (3) The set of all the people at the conference.

For simplicity we will assume that the preceding utterances of the dialogue do not add to this list of candidates, so these are all the possible options. Now crucially, the domain of discourse shifts from the first utterance, where it is something like (2) or (3), to the second and third, where it is some larger set of people. The dialogue illustrates an important fact about contexts that is sometimes overlooked: speakers have a certain degree of freedom to choose which of all the salient contexts (in this case: which of all the salient domains of discourse) they will exploit.[8] This freedom on the part of a speaker can cause problems for the hearer, who often has no *a priori* information on what domain of discourse a speaker has in mind. In such cases, it is important that both speaker and hearer can reason about what constitutes the most natural domain of discourse for each other. For example, **A** may have reasoned as follows:

> (Before **A**'s first utterance:) *We are both taking part in this conference, so it is natural for us to exploit (3) as the domain of*

[7]In what follows we will disregard the fact that, in the present example, the word *everyone* is likely to be used in a slightly vague sense that allows exceptions pretending, for simplicity, that it expresses a universal quantifier.

[8]This elusive property of contexts has recently been highlighted in a talk by Graeme Hirst at the 1997 AAAI Fall Symposium on "Context in Knowledge Representation and Natural Language".

discourse for our utterances. Thus, when I say 'everyone', this will be interpreted as quantifying over (3).

B, might have reasoned as follows:

> (Before **B**'s utterance:) ***A**'s utterance does not make sense if I take it as quantifying over a much wider set of people than (3), so it is plausible that (3) constitutes the domain of discourse for his utterance. Thus, I have to interpret 'everyone' as quantifying over (3).*

For some reason or other, however, **B** fails to reason along these lines and leaves it to **A** to clarify the resulting misunderstanding:[9]

> (Before **A**'s second utterance:) ***B**'s utterance can be explained if I assume that **B** uses a larger domain of discourse than I do. This could be either (1) or (2). In both cases, if I restrict my quantification by means of the phrase '(everyone) at this conference', **B** will understand me as expressing the proposition that I meant to express.*

It would clearly be desirable if dialogue systems were able to mimic the kind of behaviour displayed by **A**. The alternative is that both dialogue partners - i.e. the system and the user - would have to use long and fully specific Noun Phrases all the time, which would make utterances clumsy and unnatural. Note that the required reasoning, if it is correctly represented in our remarks, involves complex inferences that deal with ordinary facts (e.g. who takes part in the conference) as well as facts about contexts and this implies having a theory of the kind proposed by McCarty. In particular, **A**'s behaviour involves rules of the following kind:

> Let Dom and Dom′ be domains of discourse, where Dom $\subseteq$ Dom′. Let D be a constant denoting Dom. Then $\forall x : \phi$ is true with respect to Dom **iff** $\forall x \epsilon D : \phi$ is true with respect to Dom′.

This is a good example of what McCarthy has dubbed a lifting rule: a rule that relates sentences that are true with respect to one context (the domain Dom) to sentences that are true with respect to another context (the domain Dom′). His Ist-theory would be a natural candidate for formalizing this type of rule and it is doubtful that this can be done with equal facility in more conventional formalisms.

We have argued, in informal fashion, that understanding and generating utterances that take context into account in an intelligent way implies a

[9] Misunderstandings of this kind, and their implications for semantic theory are discussed in more detail in [van Deemter, 1998a]

style of reasoning that involves the kind of 'ontological promiscuity' where ordinary objects and contexts intermingle. The reasoning involved is not specific to the language of the utterances generated; for different languages have different mechanisms for *expressing* such notions as restricted quantification, but these notions themselves are probably universal [Barwise and Cooper, 1981]. The reasoning is precisely of the kind envisaged by McCarthy and Guha [1991], whose aim was to find ways of making the information contained in databases independent of the purpose for which the database was constructed. The reason why the Ist-theory can be applied to dialogue so smoothly is that computational databases (which originally motivated Ist-theory) and utterances by humans alike make use of abbreviated expressions, which do not get their full meaning until they are interpreted 'in context'. To us at least this suggests that there is every reason to study the potential joys of a marriage between language engineering and Ist-theory in more detail.

Kees van Deemter
ITRI, University of Brighton, UK.

Jan Odijk
Lernout & Hauspie Speech Products, Ieper, Belgium.

REFERENCES

[Barwise and Cooper, 1981] J. Barwise and R. Cooper, Generalized Quantifiers and Natural Language. *Linguistics and Philosophy* 4.

[Buvač, 1996] S. Buvač, Resolving Lexical Ambiguity using a Formal Theory of Context. In K.van Deemter and S.Peters (Eds.) Semantic Ambiguity and Underspecification. CSLI Publications, Stanford, Ca.

[Chomsky, 1981] N. Chomsky, Lectures on Government and Binding. Foris, Dordrecht.

[Chomsky, 1986] N. Chomsky, Knowledge of Language: Its Nature, Origin and Use. Praeger, New York.

[Collier and Landsbergen, 1995] R. Collier and J. Landsbergen, Language and Speech Generation. *Philips Journal of Research* 49, 1995, pp.419-437.

[Dale and Reiter, 1995] R. Dale and E. Reiter, Computational Interpretations of the Gricean Maxims in the Generation of referring Expressions. *Cognitive Science* 18, p.233-263

[van Deemter and Odijk, 1997] K. van Deemter and J. Odijk. Context Modeling and the Generation of Spoken Discourse. *Speech Communication* 21, pp.101-121.

[van Deemter, 1998a] K. van Deemter. Domains of Discourse and the Semantics of Ambiguous Utterances: a Reply to Gauker. *Mind* 107, No.426, April 1998.

[van Deemter, 1998b] K. van Deemter. A Blackboard Model of Accenting. *Computer Speech and Language* 12 (3), June 1998.

[Demolombe and Jones, 1995] Reasoning about Topics: Towards a Formal Theory. In Working Notes of the Workshop on Formalizing Context, AAAI-1995 Fall Symposium Series.

[Dirksen, 1992] A. Dirksen, Accenting and deaccenting: a declarative approach. Proc. of Coling Conference 1992, Nantes, France.

[Dorrepaal, 1990] J. Dorrepaal, Discourse
anaphora. In Proceedings of COLING 1990 conference, vol II, p. 95-99.

[Grosz *et al.*, 1986] B. J. Grosz, K. S. Jones and B. L. Webber, 'Readings in Natural Language Processing', Morgan Kaufman Publishers, Los Altos, Cal.

[Grosz and Sidner, 1986] B. J. Grosz and C. L. Sidner. Attentions, Intentions and the Structure of Discourse. *Computational Linguistics*, 12, 175-204.

[Guha, 1991] R. V. Guha, Contexts: A Formalization and some Applications. PhD thesis, Stanford University.

[Halliday, 1967] M. A. K. Halliday, Notes on transitivity and theme in English. *Journal of Linguistics* 3: 199-244.

['t Hart *et al.*, 1990] J. 't Hart, R. Collier and A. Cohen, A Perceptual Study of Intonation. Cambridge University Press, Cambridge.

[Hirschberg, 1990] J. Hirschberg, Accent and discourse context: assigning pitch accent in synthetic speech. Proc. of AAAI 1990, p.953.

[Kamp and Reyle, 1994] H. Kamp and U. Reyle, From Discourse to Logic. Kluwer, Dordrecht.

[Ladd, 1980] D. R. Ladd, The structure of intonational meaning: evidence from English. Indiana Univ. Press, Bloomington, In.

[McCarthy, 1993] J. McCarthy, Notes on Formalizing Context. In Proc. of 13th IJCAI conf.

[Odijk 1994] J. Odijk, Text generation without planning. Paper presented at CLIN V, University of Twente.

[Odijk, 1995] J. Odijk, Generation of Coherent Monologues, in T. Andernach et al. (eds.), CLIN V: Proceedings of the Fifth CLIN Meeting, pp. 123-131, University of Enschedé, The Netherlands.

[Prevost and Steedman, 1994] S. Prevost and M. Steedman, Specifying Intonation from Context for Speech Synthesis. *Speech Communication.*

[Rosetta, 1994] M. T. Rosetta, Compositional Translation, Kluwer Academic Publishers, Dordrecht.

[Terken and Nooteboom, 1987] J. Terken and S. Nooteboom, Opposite effects of accentuation and deaccentuation on verification latencies for 'given' and 'new' information. Language and Cognitive Processes 2, 3/4, pp.145-63.

[Willems *et al.*, 1988] N. Willems, R. Collier and J. 't Hart, A synthesis scheme for British English intonation. *J. Acoust. Soc. Am.* 84 (4).

HARRY BUNT

REQUIREMENTS FOR DIALOGUE CONTEXT MODELLING

1 INTRODUCTION

This paper is concerned with context in relation to language understanding, focusing on the interactive use of language in dialogues. Language understanding in general requires contextual information in order to assign appropriate meanings to linguistic expressions; the understanding of utterances in dialogue involves context in an additional, more fundamental way, in that it requires us to relate utterances to their *causes* and *intended effects*. In other words, for language used in interaction, the very notions of *understanding* and *utterance meaning* should be defined in terms of context properties and how they change through communication.

According to this view, the analysis of utterance meaning in dialogue can provide insights into the conceptual content of the contexts changed by these utterances. More specifically, an analysis of the different functions of utterances in dialogue and their intended context-changing effects tells us what kinds of information such a notion of context should include, and allows us to deduce certain formal properties of that information, thus providing clues for the formal and computational modelling of contexts.

The study of utterance meanings as intended context-changing actions tells us why utterances occur, while on the other hand the study of what motivates communicative activity gives us information about the components of utterance meanings. We therefore believe that the study of utterance meanings and dialogue mechanisms is most fruitfully pursued in combination, within a single theoretical framework. Starting with Bunt [1989], we have been developing such a framework, called *Dynamic Interpretation Theory.* In order to make this approach work, the notion of context must be given a manageable interpretation. To this end we investigate communicative activity in dialogues and analyse the aspects of context addressed by communicative acts.

This paper is organized as follows. We first give a brief summary of the main features and assumptions of Dynamic Interpretation Theory as far as relevant to the purpose of this paper. We then turn to the analysis of different classes of communicative acts and the requirements they pose on context modelling. We finally discuss the conceptual dimensions of the relevant notion of context and suggest possible directions for formal and computational context modelling.

2 DYNAMIC INTERPRETATION THEORY

Dynamic Interpretation Theory (DIT) has emerged from the study of spoken human-human information dialogues, and aims at uncovering fundamental principles in

P. Bonzon, M. Cavalcanti and R. Nossum (eds.), Formal Aspects of Context, 23–36.

dialogue both for the purpose of understanding natural dialogue phenomena and for designing effective, efficient and pleasant computer dialogue systems. Information dialogues serve the purpose of exchanging factual information concerning some domain of discourse. Such a task naturally gives rise to questions, answers, checks, confirmations, etc. In addition, information dialogues also contain other elements such as greetings, apologies, and acknowledgements. We refer to the first type of elements as *task-oriented* acts and to the latter as *dialogue control* acts. Task-oriented acts are directly motivated by the task or purpose motivating the dialogue and contribute to its accomplishment; dialogue control acts are concerned with the interaction itself, and serve to create and maintain the conditions for smooth and successful communication.

2.1 Dialogue acts

The idea that linguistic behaviour may be understood in terms of context changes is closely related to action-based views of language, of which speech act theory is the most prominent representative. In fact, the two views may coincide, when we view communicative actions as aimed at changing the context. An important technical difference between DIT and speech act theory is that a context-change approach takes context and context changes as most fundamental notions, where speech act theories tend to take illocutionary acts as the fundamental concepts (see Bunt, 1998 for more about the difference between DIT and other action-based approaches). To describe the context-changing effects of dialogue utterances, we have introduced the concept of a *dialogue act,* defined as the functional units used by the speaker to change the context [Bunt, 1994]. A dialogue act has a semantic content, formed by the information the speaker introduces into the context, and a communicative function that defines the significance of this information by specifying how the context should be updated with the information. The stipulation ... *used by speakers* in the definition of a dialogue act ensures that every communicative function corresponds to a particular set of features of observable communicative behaviour. Formally, a communicative function is a mathematical function that, given a semantic content as argument, maps a context into a new context.

In Bunt [1989] we have presented a hierarchical system of task-oriented dialogue acts defined as context-changing operations, where context is construed as the pair consisting of the states of information of the two participants. At the top of this hierarchy we find two subclasses of dialogue acts, those concerned with *information seeking* and those with *information providing.* For dialogue control acts a classification into three subsystems, concerned with *feedback, interaction management,* and *social obligation management* has been proposed [Bunt, 1994]. Feedback acts provide information about the processing of inputs, reporting or resolving problems (negative feedback), or reporting successful processing (positive feedback). Feedback may relate to the speaker's own processing ('auto-feedback') or to the other agent's processing ('allo-feedback'). Interaction management acts handle various aspects of the interactive situation, such as taking turns, pausing and resuming, and monitoring attention and contact. Social obligation management acts deal with socially indicated obligations such as welcome greeting, thanking, apologizing, and farewell greeting.

For task-oriented dialogue acts the communicative function may be one that is specific to a particular type of discourse situation, or it may be a function that can be used in any type of dialogue. For instance, in negotiation dialogues one may find functions that are rather specific to that situation, such as OFFER and REFUSAL (see [Alexandersson, 1996]). On the other hand, we find questions, answers, informs, verification, confirmations, etc. in virtually any type of dialogue. In a negotiation dialogue we may for example make an offer by using the OFFER function, as in *'How about Friday the 13th?'* (assuming that *'How about X'* is a special form for making offers), but alternatively we can use an INFORM function to make an indirect offer, and say: *'Friday the 13th would be fine with me'*. For dialogue control acts the situation is similar: there are communicative functions specific for dialogue control purposes, corresponding to the use of special-purpose utterance forms, but the general devices for informing, questioning, confirming, etc. may also be used to construct a dialogue control act. An example is *'Thank you'*, using the THANKING function, in contrast with *'I am very grateful to you'*, using an INFORM function.

Figure 1 provides a schematic overview of the subsystems of communicative functions we have identified, where task-specific communicative functions can be used only to build a task-oriented dialogue act; dialogue control functions can be used only to build a dialogue control act, and informative functions can be used to build either kind of dialogue act, depending on the semantic content. Notice that the 'task-specific' functions in an information dialogue would be just the information-seeking and -providing functions; therefore, information dialogues constitute the one and only kind of dialogue for which there are in fact *no* task-specific communicative functions.

2.2 Dimensions of context

Studies of meaning in discourse often acknowledge the crucial role of context, but explicit modelling of context is nearly always restricted to only a few aspects of context. Studies of texts as discourses tend to model only the linguistic ('discourse') context, e.g. in discourse representation structures [Kamp and Reyle, 1993]. Studies of dialogue, on the other hand, tend to focus on the 'cognitive' context as constituted by the participants' beliefs, desires, and intentions, as in BDI-type models [Bratman *et al.*, 1988]; see also [Traum, 1996; Beun, 1996], or on participants' intention and attention [Grosz and Sidner, 1986]; see also [Vilant *et al.*, 1994; Rats, 1996; Maier, 1995; Maier, 1996]. To account for the meanings of all the types of dialogue utterances that one finds in dialogue, we believe that a broader notion of context should be modelled.

As mentioned above, utterance meaning is defined in DIT in terms of context

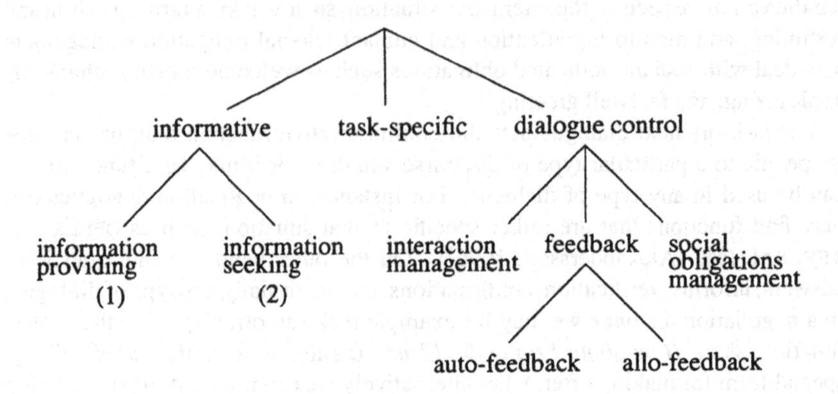

(1), (2): hierarchies of communicative functions defined in Bunt (1989).

Figure 1. Subsystems of communicative functions

change. In the literature, the term 'context' is used in many different ways, referring for example to the preceding discourse, to the physical environment, or to the domain of discourse. In Bunt [1994] we have argued that these factors can be grouped into five categories: *cognitive, semantic, physical, social,* and *linguistic*. For each of these 'dimensions' of context we distinguish between *global* aspects, which are constant throughout the dialogue, and *local* aspects, whose values change during and, more specifically, *through* the dialogue. These categories of context factors may be characterized briefly as follows.

- **Linguistic context:** surrounding linguistic material, 'raw' as well as analysed. Closely related to what is sometimes called 'Dialogue History' (see e.g. [Bilange, 1991; Prince and Pernel, 1995; Bunt, 1998]).
- **Semantic context:** state of the underlying task; facts in the task domain.
- **Cognitive context:** participants' states of processing and models of each other's states.
- **Physical and perceptual context:** availability of communicative and perceptual channels; partners' presence and attention.
- **Social context:** the participants' communicative rights, obligations and constraints.

Every dialogue act, when processed by the addressee, changes the addressee's cognitive and linguistic context; the difference between a task-oriented dialogue act and a dialogue control act is that the former is aimed at changing the semantic context (the state of the underlying task), whereas a dialogue control act is intended to change the social, cognitive, or physical and perceptual context.

2.3 Why communicative agents act

In order to be able to develop an explanatory theory of dialogue, DIT makes the following idealizing assumptions about human communicative agents:

Rationality People communicate in order to achieve something. They form communicative goals in accordance with underlying goals and desires, choose their communicative actions so as to further their communicative goals, and organize interaction so as to optimize the chances of success of their communicative actions.

Sociality Communication between people is a form of social behaviour, and is thus subject to cultural norms and conventions for interaction. An important aspect of this is **Cooperativity**, i.e. taking the dialogue partner's goals, limitations, and other characteristics into account in the choice of one's communicative actions and the way they are expressed.

From these basic assumptions, the motivations of the various types of communicative act can be derived as follows.

- Task-oriented acts: acts with a task-specific function, as well as those with an informative function and a task-related semantic content, are motivated by the speaker's underlying task (Rationality) or by his knowledge of the partner's task (Cooperativity).
- Dialogue control: *Positive feedback* acts are motivated by a social principle, saying that feedback is required from time to time (Sociality). *Negative feedback* is motivated by the drive to communicate successfully; this requires any obstacles in this respect to be removed (Rationality). Similarly for *interaction management* acts. *Social obligation management* acts are motivated by the desire to honour social obligations (Sociality).

3 CONTEXT MODELLING

Communicative action, as opposed to physical action, cannot change anything in the physical world, but only something in the 'mental worlds' of the communicating agents. In Bunt [1989] we construed local context models for task-oriented dialogue acts as pairs $< K_A, K_B >$, where K_A is the state of knowledge, belief and intentions regarding the discourse domain of partner A, and K_B that of B, and where both K_A and K_B include the mutual beliefs of the respective agents vis-à-vis each other. This approach has been implemented in a primitive way in the TENDUM dialogue system (see [Bunt *et al.*, 1984]), and is currently being implemented in a type-theoretical formalism in the DENK system [Ahn *et al.*, 1994; Bunt *et al.*, 1998a; Bunt *et al.*, 1998b]. This approach to local context seems adequate for dealing with task-oriented acts, but not for dialogue control acts. In Bunt

[1995] we therefore proposed a richer notion of context, where local context models are pairs $< C_A, C_B >$, C_A being the local context according to A, consisting of the five dimensions mentioned above, and C_B that according to B. Within an agent's local cognitive context, moreover, two components are distinguished: the speaker's own processing status and his model of the hearer's model of the local context. This latter part makes the structure recursive; this is required, since any time the speaker performs a dialogue act concerning some aspect of local context he displays an assumption about the hearer's corresponding type of context information. Using C_{AB} for the local context as A believes B views it, we thus have for A's model of local context the structure depicted in Figure 2.

C_A = < A's local semantic context,
A's local cognitive context: < Own Processing Status
Partner Model: C_{AB} >
A's local physical and perceptual context,
A's local social context,
A's local linguistic context >

Figure 2. Conceptual structure of an agent's local context

3.1 Task-oriented dialogue acts

Task-oriented dialogue acts ('TO-acts') are motivated by a communicative goal that derives from an underlying noncommunicative, task-specific goal, and are assumed to be the product of processes of rational deliberation, such as inferencing and planning. The understanding of such an act creates in the hearer the belief that the speaker's state of intention and information has the properties expressed by the semantic content and the communicative function. A TO-act thus changes the hearer's local cognitive context, as do all dialogue acts, and it aims at changing the semantic context.

It seems fairly obvious that the most complex kind of local context information is formed by the participants' beliefs and intentions w.r.t. to the underlying task (semantic context), plus their beliefs and communicative intentions w.r.t. each other's task-related beliefs and intentions - and this recursively. It is this information that is expressed in articulate semantic contents in TO-acts, and that may be the subject of elaborate discussions in a dialogue; clearly, this is also what agents reason about and form intentions about. This information therefore requires an expressive representation formalism with associated inference machinery. For the generation of dialogue acts in an information dialogue, an agent's most important kinds of belief (and intention) are thus:

1. his beliefs about (and intentions regarding) the state of the task;

2. his beliefs about (and intentions regarding) the partner's beliefs about (and intentions regarding) the state of the task;

3. his beliefs about (and intentions regarding) the mutual beliefs about the state of the task.

Two promising formalisms for representing such information are *Constructive Type Theory* (see [Ahn *et al.*, 1994; Kievit, 1998; Piwek, 1998] and *Modular Partial Models* (see [Bunt, 1998]). The first of these is currently being implemented in the DENK multimodal dialogue project (see [Bunt *et al.*, 1998a; Bunt *et al.*, 1998b]), in a way that focuses on the representation of the items 1 and 3; the second one has been implemented in a limited fashion in the ΔELTA dialogue project [Bunt, 1995], focusing on the representation of finite nestings of beliefs, like those of the types 1 and 2.

3.2 Dialogue control acts

Dialogue control acts ('DC-acts') with DC-functions typically have no or only marginal semantic content; their meaning is concentrated in their communicative function. Since no articulate semantic content is involved, we conjecture that the generation of such acts does not involve reasoning with representations of complex beliefs and intentions, but that these acts are triggered by relatively simple conditions that can be expressed in terms of the values of a small number of parameters.

Feedback acts

Feedback acts are triggered by difficulties that the speaker encounters in processing an incoming utterance (negative feedback), or by successful completion of such processing (positive feedback). In performing feedback acts, speakers show that they have knowledge of the difficulties as well as of the successes of their own processing, and that they also have beliefs about their partner's processing. This is the information that we call *'processing status'*. A speaker's knowledge of his own processing forms part of his local cognitive context; his beliefs about the hearer's processing forms part of the recursively embedded believed hearer's model of the local context.

Social obligations management acts

In natural communication there are certain things one is supposed to do and certain things one is supposed not to do, following general norms and conventions of social behaviour in the culture to which one belongs. When contacting someone with the purpose of engaging in a dialogue, one may e.g. be supposed to exchange greetings, and if one doesn't know the other agent, one is supposed to introduce oneself. We use he term 'social obligations' to describe such situations.

For dealing with social obligations, languages have closed classes of utterance types with the special property that using an utterance of such a class puts a pressure on the addressee to react using a particular type of utterance from a related (sometimes overlapping) closed class. For example, a *'Good morning Mary!'* creates a pressure to respond with something like *'Good morning John!'*, and *'Thank you very much'* to respond with *'You're welcome'*. In Bunt [1994] we have introduced the notion of *reactive pressures* (RPs) to capture this phenomenon. In our information dialogue corpora we have found five types of situation where social obligations are dealt with: those where greetings are in order (beginning and end of the interaction); those where agents introduce themselves; those where an agent apologizes for a mistake or for the inability to supply requested information, and those where gratitude is expressed.

A closed class-expression introducing a reactive pressure corresponds to what in the terminology of the Geneva School is called an 'initiative' act, and the utterance that the agent is pressured to perform corresponds to a 'reaction' element (see [Moeschler, 1985; Roulet, 1985]). In order to account for the occurrence of initiative utterances dealing with social obligations, we have introduced the concept of *interactive pressures* [Bunt, 1995]. An interactive pressure (IP), like a reactive pressure, is a pressure on an agent to perform a certain communicative action; the difference with an RP is that it is created not by a particular utterance, but by properties of the local context. IPs lead to initiative dialogue acts for social obligation management if we assume that communicative agents tend to act to resolve such pressures, thereby evoking the corresponding reactive acts through the RPs of the initiative ones.

An IP is created by an IP principle, which is a pair consisting of (1) a set of local context conditions that must be satisfied for the pressure to arise; (2) a specification of properties of communicative acts that would resolve the pressure, notably a communicative function (or class of functions) and a set of constraints on semantic content and utterance form. (See [Bunt, 1996] for more about IP principles.)

Whenever the conditions in an IP principle are satisfied, this leads to a partly specified dialogue act to become 'active' in the local social context. When more than one act is active in local social context, the last act that became active usually seems to have priority to be performed first or exclusively, as in *'Airport Information Service, good morning'/'Good morning, this is Jansen'* and *'Thanks a lot, goodbye'/'Goodbye'*. This suggests that a stack is an appropriate organization of the local social context, with decreasing strength of the interactive/reactive pressure on elements lower on the stack.

Interaction management acts

Interaction management (IM) acts form a rather heterogeneous class of dialogue acts, which in the case of spoken information dialogues are primarily concerned with turn-taking, timing, contact, dialogue structuring, and the utterance formulation process. In a corpus of 111 naturally occurring spoken information dialogues (see [Beun, 1989]) the most important cases of turn management were the following:

1. The information service encourages the client to continue (TURN-GIVING).
2. The speaker (information service or client) needs a little time for producing a response, but wants to keep the turn (TURN-KEEPING).
3. The present speaker is interrupted, because the hearer detects an error (INTERRUPTION).

At any time in a dialogue, there is normally one participant who has the main speaker role, and both participants have a view on who is having that role, is 'having the turn'. The beliefs that participants have about this are best treated as integrated with the linguistic context, which records the linguistic events that take place, and can thus be seen as constituting a 'dialogue history' (cf. [Prince and Pernel, 1995]). This recorded dialogue history includes a representation of the turn-taking in the preceding dialogue. The linguistic context is not purely backward looking, however: in order to take the dialogue participants' anticipations as to how the dialogue will continue into account, a planned or expected *'dialogue future'* is also needed, which contains the same kind of information as the dialogue history, but with less detail. The present and future turn-taking situation can be represented simply by a parameter indicating the speaker of each contribution. This information is sufficient for accounting for TURN-GIVING and TURN-KEEPING acts. The conditions motivating an INTERRUPTION are already represented in the partner's processing status.

Time management acts are concerned with activities that the speaker is planning. A PAUSE occurs when the speaker estimates that the time he needs is too long for an unexplained silence. When he thinks relatively little time is needed, he may instead STALL (speaking more slowly, producing *'ehm'*s etc.). To account for time management acts, we must assume that speakers have estimates of the time needed to perform or to complete input processing, further cognitive and task-specific processing, and output generation. We add this to the processing status information in the local cognitive context of each participant.

Time management acts typically indicate only that *some* extra time is needed, not how much; at most, they specify the activity for which time is needed (*'One moment please, I will look that up for you'*). The description of activities in these messages is not a new type of information; it also occurs in explicit feedback acts, the additional aspect entering here being the time estimated for (completing) an activity. Since this time estimate is never articulate (specific requests for time like *'Twenty-five seconds please'* are never found), time management requires only one extra parameter per activity (per 'process') in the representation of processing status, with only 3 possible values: 'negligible', 'small' (for STALLING), and 'substantial' (for PAUSE).

Contact management acts frequently occur in an explicit form in telephone dialogues, when the speaker is uncertain whether the person at the other end of the line is actually there and is paying attention. Such uncertainty may occur especially after a pause. These acts require the speaker to make assumptions about the physical and mental 'presence' of his dialogue partner. To represent this information, two parameters are sufficient: one for the speaker's presence & attention and one for that of the partner. These parameters are part of the local physical/perceptual context representation.

In our corpus investigations, we have identified two types of dialogue acts concerned with the utterance formulation process: RETRACTION, where the speaker has made an error and retracts a contribution or part of it, and SELF-CORRECTION, in which case the speaker replaces some erroneously produced material by something else. These acts do not require any new kind of context information to be modelled beyond that which is already needed for feedback purposes; the only extension required is that output generation processes should be among the processes whose status is represented in the processing status part of the local cognitive context.

Discourse structuring acts, finally, are performed by a speaker in order to structure the interaction, indicating for example that he is closing the discussion of a certain topic, that he wants to address a new topic, or that he wants to ask a question. Discourse structuring acts are based on the speaker's view of the current linguistic context and on his plan for continuing the dialogue. Topic management requires certain elements in the linguistic context to be identifiable as having topical status. Our corpus investigations provide little evidence of elaborate communicative planning in information dialogues, as might have been reflected in topic management acts (cf. [Rats, 1996; Rats and Bunt, 1997]). As far as the planning of topics is concerned, speakers in information dialogues do not seem to do more then (1) deciding on a set of topics to be addressed; (2) selecting the next topic from this set as the dialogue continues. The representation of complex topical structures therefore does not seem necessary. For the generation of dialogue structuring acts, it is sufficient to have an articulate representation of goals and related aspects of the state of information, especially if the linguistic context keeps a record of goals that have been established earlier in the dialogue and that have been achieved, suspended, or abandoned. We thus assume that the information needed for discourse structuring acts can be found in the cognitive and linguistic context, rather than in a separate part of the context model.

Feedback, utterance formulation, and time management require a representation of process information that can be realized by means of a small number of parameters per process, as shown in attribute-value matrix form in Figure 3. The RESULT parameter is intended to have a list of values R_i which may be complex structures, like nested feature structures.

Process P:	
PROGRESS:	ended (/ongoing/suspended/planned)
RESULT:	$< R_1, R_2, .., R_k >$
RESULT-TYPE:	complete (/partial/zero)
LACK-INFO:	none (/...)
TIME-NEED:	negligible (/small/substantial)

Figure 3. Processing status information represented in local cognitive context

It might be objected that the structure, depicted in Figure 3, leads to too much nesting for IM information. Indeed, turn-taking, timing, self-correction, contact, and discourse structure are never the subject of explicit communication, consistent with the fact that these acts mostly have a marginal propositional content. It therefore seems implausible that dialogue participants would build elaborate recursive structures for this kind of information. An attractive alternative may therefore be to restrict the recursion to beliefs relating to the semantic context, while the other components have a limited, finite nesting of only a few levels deep. (See further [Bunt, 1998].)

3.3 *Overall aspects of context representation*

Using a simple parameter-based representation for IM-information, as in Figure 3, greatly simplifies the implementation of a context model to be used for effective and efficient interaction management. It should on the other hand be acknowledged that dialogue participants sometimes do form nested and quantified beliefs about IM conditions; for instance, when partner I says *'Een ogenblikje'* (*'Just a moment'*), partner C knows that I intends to execute a process P that requires substantial time. We can represent this in a formal language with facilities for expressing beliefs by construing the use of parameter specifications as a way to form predicates. Using B_C and B_I for belief operators relating to the subscripted partners, we thus want to represent something like:

$$B_C((\exists P : B_I([\text{TIME-NEEDED: } subst](P)))$$

(C believes that there is a process P that I believes requires substantial time.) This representation is based on the assumption that a parameter specification, like [TIME-NEEDED: *subst*], can be considered as a feature specification, and thus as equivalent to a predicate, like $(\lambda x : \text{TIME-NEEDED}(x) = subst)$. Feature structures can be formally integrated in logical languages, as has been shown in [Bunt and van der Sloot, 1996]; this technique can be used here to integrate a simple parameter-based representation of IM-information in the expressive representation language required for task-oriented semantic information. Such an integration is needed for dealing with IM-acts which are expressed not by using a specific IM-utterance form, but by using a full sentence, as in *'I will need some time to find that information in the new time table, valid from this Monday'*. We have not encountered such cases in our corpus, but they are obviously possible, and their treatment should not be excluded by relying entirely on a simple parameter-based representation. We can handle this by allowing feature attributes to have values consisting of complex expressions in a logical language.

Parameter- or feature-based representations are attractive not only for their conceptual and computational simplicity, but also because they allow the expression of constraints by means of *structure sharing*. This seems a useful property in context modelling, since there may be desirable dependencies between elements in different context dimensions for two reasons:

- The local linguistic context, which we did not discuss in this paper in any detail (see [Bunt, 1998]) naturally acts as a kind of dialogue history, representing the communicative events in the dialogue with their syntactic, semantic and pragmatic analysis. Elements of these analyses are bound to reappear in other context dimensions, and it may be useful to link the various occurrences in order to be able to reconstruct a previous context when necessary, for instance when a misunderstanding is addressed.

- It is convenient to allow the same object to appear in different context dimensions, because it may be significant from different points of view. For instance, when a subdialogue is concerned with a topic *t*, it is also the case that one of the participants has a goal involving *t*, and one also expects to find utterances with the utterance topic *t*. The notions 'goal', 'topic', and 'dialogue structure' thus typically all involve the same object; structure sharing may be used to represent this, and thus to exploit for example the interpretation of a topic management act in recognizing the speaker's intention in a subsequent TO-question.

4 CONCLUSIONS

In this paper we have explored the notion of *local context*, needed for describing the meanings of utterances in dialogue. We showed that the analysis of utterance meanings can provide insights into the kinds of elements that this notion of context must include, as well as into the formal properties of context elements and into possibilities of formal and computational modelling. The work described in this paper is ongoing research, and the conclusions about dialogue context modelling that we can draw at this point require further analysis and testing. In summary, our conclusion is that, conceptually, dialogue context should contain the five kinds of information indicated in Fig. 2, and that the representation of this information involves four kinds of structures:

1. Representation structures for beliefs and intentions regarding the task, regarding the partner's beliefs and intentions, and regarding mutual beliefs (local semantic and cognitive context). We have suggested Constructive Type Theory and Modular Partial Models as candidate formalisms for these structures.

2. Representations of past and planned communicative events as dialogue acts, making up the local linguistic context (not addressed in this paper; see [Bunt, 1998]). Recursive feature structures combined with a formalism for representing underspecified logical forms appear to be adequate.

3. A stack of partial specifications of dialogue acts for representing the local social context.

4. Simple attribute-value matrices for representing the local physical and perceptual context and the processing status information in the local cognitive context.

Finally, we have concluded that it would be useful to define a single representation formalism (with structure sharing or other linking facilities) incorporating all four kinds of structure, in order to represent logical relations between elements in the different dimensions of context.

Tilburg University, The Netherlands.

REFERENCES

[Ahn *et al.*, 1994] R. Ahn, R. J. Beun, T. Borghuis, H. C. Bunt and C. van Overveld. The DENK architecture: a fundamental approach to user interfaces. *Artificial Intelligence Review*, **8**, 431–445, 1994.

[Alexandersson, 1996] J. Alexandersson. Some ideas for the automatic acquisition of dialogue structure. In *Proc. 11th Twente Workshop on Language Technology.* Enschede: University of Twente, pp. 149–158, 1996.

[Beun, 1989] R. J. Beun. *The recognition of declarative questions in information dialogues.* PhD Thesis, Tilburg University, 1989.

[Beun, 1996] R. J. Beun. Speech act generation in cooperative dialogue. In *Proc. 11th Twente Workshop on Language Technology.* Enschede: University of Twente, pp. 71–79, 1996.

[Bilange, 1991] E. Bilange. *Modélisation du dialogue oral personne-machine par une approche structurelle: theorie et réalisation.* PhD Thesis, Rennes 1 University, 1991.

[Bratman *et al.*, 1988] M. E. Bratman, D. J. Israel and M. E. Pollack. Plans and resource-bounded practical reasoning. *Computational Intelligence*, **4**, 349–355, 1988.

[Bunt, 1989] H. C. Bunt. Information Dialogues as Communicative Actions in Relation to User Modelling and Information Processing. In *The Structure of Multimodal Dialogue.* M. M. Taylor, F. Néel and D. G. Bouwhuis, eds. pp. 47–73. North-Holland Elsevier, Amsterdam, 1989.

[Bunt, 1994] H. C. Bunt. Context and Dialogue Control. *THINK Quarterly*, **3**, 19–31, 1994.

[Bunt, 1995] H. C. Bunt. Semantics and pragmatics in the ΔELTA dialogue system. In *Proc. of the 2nd Spoken Dialogue and Discourse Workshop.* L. Dybkjaer, ed. pp. 1–27, Roskilde University, 1995.

[Bunt, 1996] H. C. Bunt. Dynamic Interpretation and Dialogue Theory. In *The Structure of Multimodal Dialogue, Vol. 2.*, M. M. Taylor, F. Néel and D. G. Bouwhuis, eds. John Benjamins, Amsterdam, 1996.

[Bunt, 1998] H. C. Bunt. Iterative context specification and dialogue analysis. In *Abduction, Belief and Context: Studies in Computational Pragmatics.* H. C. Bunt and W. J. Black, eds. pp. 73–129. University College Press, London, 1998.

[Bunt *et al.*, 1984] H. C. Bunt, R. J. Beun, Dols, J. van der Linden and G. Schwartzenberg. The TENDUM dialogue system and its theoretical basis. *IPO Annual Progress Report,* **19,** 105–113, 1984.

[Bunt and van der Sloot, 1996] H. C. Bunt and K. van der Sloot. Parsing as dynamic interpretation of feature structures. In *Recent Advances in Parsing Technology.* H. Bunt and M. Tomita, eds. Kluwer Academic Press, Dordrecht, 1996.

[Bunt *et al.*, 1998a] H. C. Bunt, R. Ahn, L. Kievit, P. Piwek, M. Verlinden, R. J. Beun, and C. van Overveld. Multimodal Cooperation with the DENK System. In *Multimodal Human-Computer Communication: Systems, Techniques and Experiments.* H. C. Bunt, R. J. Beun and T. Borghuis, eds. pp. 39–67. Lecture Notes in Artificial Intelligence 1374, Springer-Verlag, Berlin, 1998.

[Bunt *et al.*, 1998b] H. C. Bunt, L. Kievit, P. Piwek, M. Verlinden, R. J. Beun, T. Borghuis and C. van Overveld. Cooperative dialogue with the multimodal DENK system. In *Proc. 2nd Int. Conference of Cooperative Multimodal Communication CMC/98.* H. C. Bunt, R. J. Beun, T. Borghuis, L. Kievit and M. Verlinden, eds. SOBU, Tilburg University, 1998.

[Grosz and Sidner, 1986] B. J. Grosz and C. L. Sidner. Attention, intention, and the structure of discourse. *Computational Linguistics,* **12,** 175–204, 1986.

[Kamp and Reyle, 1993] H. Kamp and U. Reyle. *From Discourse to Logic.* Kluwer Academic Press, Dordrecht, 1993.

[Kievit, 1998] L. Kievit. *Context-driven natural language interpretation.* PhD Thesis, Tilburg University, 1998.

[Maier, 1995] E. Maier. A multi-dimensional representation of context in a speech translation system. In *Proc. IJCAI'95 Workshop on Context in Natural Language Processing,* L. Iwanka and W. Zadrozny, eds. pp. 78–85, 1995.

[Maier, 1996] E. Maier. Context construction as a subtask of dialogue processing - the Verbmobil case. In *Proc. 11th Twente Workshop on Language Technology.* pp. 113–122. University of Twente, Enschede, 1996.

[Moeschler, 1985] J. Moeschler. *Argumentation et conversation, éléments pour une analyse pragmatique du discours.* Hatier, Paris, 1985.

[Piwek, 1998] P. Piwek. *Logic, Information & Conversation.* PhD Thesis, Eindhoven University of Technology, 1998.

[Prince and Pernel, 1995] V. Prince and D. Pernel. Several knowledge models and a blackboard memory for human-machine robust dialogues. *Natural Language Engineering,* **1,** 113–145, 1995.

[Rats, 1996] M. Rats. *Topic management in information dialogues.* PhD Thesis, Tilburg University, 1996.

[Rats and Bunt, 1997] M. Rats and H. C. Bunt. Information Packaging in Dutch Information Dialogues. In *Proc. 3rd Spoken Dialogue and Discourse Workshop.* L. Dybkjaer, ed. pp. 52–73. Maersk Mc-Kinney Moller Institute for Production Technology, Odense, 1997.

[Roulet, 1985] E. Roulet. *L'articulation du discours en francais contemporain.* Peter Lang, Bern, 1985.

[Searle, 1969] J. R. Searle *Speech Act Theory.* Cambridge University Press, Cambridge, 1969.

[Traum, 1996] D. R. Traum. The TRAINS-93 Dialogue Manager. In *Proc. 11th Twente Workshop on Language Technology,* pp. 1–11. University of Twente, Enschede, 1996.

[Vilant *et al.*, 1994] A. Vilnat, B. Grau and G. Sabah) Control in man-machine dialogue. *THINK,* **3,** 32–55, 1994.

JULIA LAVID

CONTEXTUAL CONSTRAINTS ON THEMATIZATION IN WRITTEN DISCOURSE: AN EMPIRICAL STUDY

1 INTRODUCTION

Context modelling and its influence on the linguistic structure of texts has been the object of much research, especially within functional theories of language [Halliday, 1978; Kress and Threadgold, 1988; Lemke, 1988; Fairclough, 1989]. In these theories the issue is how broadly the concept of context is defined. For earlier functional linguists, it referred basically to the linguistic context, whereas for some contemporary theorists who operate within a much broader social semiotics, context needs to be more discursively understood as a multi-levelled phenomenon, and 'text' as the product of varying contextual levels and components. In general, researchers have recognized at least three levels of context—cultural, situational and textual—each one consisting of different components and variables, with variations in the schemes of classifications, and in interpretations of the significance of the concepts.[1]

The concept of *theme* and its development has also been the object of much research in the last few years, especially among systemic-functional linguists [Ghadessy, 1995]. Although most scholars have investigated its meaning as a functional unit within the structure of the clause [Halliday, 1967; Halliday, 1985; Firbas, 1992; Fries, 1981; Martin, 1995], recent studies have also paid attention to the discourse functions of thematic information and its relevance to the description of texts [Fries, 1981]. More specifically, Fries [1995, p. 10] hypothesized a relationship between theme selection and genre type, and different authors explored this relationship in different genres/registers [Berry, 1989; Bäcklund, 1990; Francis, 1989; Francis, 1990; Ghadessy, 1995; Wang, 1991]. The results obtained in these studies varied greatly depending on the genre chosen, partly due to the lack of rigorous criteria for genre assignment, and partly due to the limited data used for analysis.

This study presents an investigation of the relationship between two contextual variables, widely recognized as such in different models of context [Halliday, 1978; Hymes, 1974; Rubin, 1984], and the phenomenon of thematization in discourse. These two contextual variables are the *discourse purpose* and the *subject matter* of the text. The relationship was explored in two phases:

[1]For a revision of some of the most important classification schemes see Leckie-Tarry [1995, pp. 17–30].

P. Bonzon, M. Cavalcanti and R. Nossum (eds.), Formal Aspects of Context, 37–47.

a) on a first phase of analysis, an analysis was carried out of the distribution of the organizational patterns (chaining strategies) which appear in different text-type groups characterized by specific combinations of the contextual variables specified above;

b) on a second phase of analysis, the analysis concentrated on the semantic types of *themes* selected to signal a given chaining strategy in the previously established text-type groups and their correlations with the markers of the different possible strategies.

The paper is organized as follows: Section 2 concentrates on the corpus, and explains the methodology used for its analysis. Section 3 presents the results, and Section 4 discusses the main findings of the empirical analysis and some implications for automated text generation.

2 MATERIALS AND METHOD

Materials

A corpus of a hundred texts was used for analysis. The sample was collected as follows: texts belonging to different text types and genres, according to the typology established by [Werlich, 1983], were distributed by three independent judges in five different groups characterized by specific combinations of the contextual variables of discourse purpose (DP) and subject matter (SM).[2] In this study, the variable of *discourse purpose* was operationalized as the overall communicative goal that the text producer has in mind, while the *subject matter* is the abstract propositional content of the text. The discourse purposes used in this study correspond to the well-known discourse types found in several existing text typologies, e.g. narrative, expository, descriptive, instructive and argumentative (see [Werlich, 1983]). The five text-type groupings were the following:

- Group 1: this group consisted of twenty texts characterized by an *expository* discourse purpose and a subject matter dealing with *whole classes of objects* and/or *generic concepts*. The texts included in this group were basically encyclopedia entries.

- Group 2: this group consisted of twenty texts characterized by a *descriptive* discourse purpose and a subject matter dealing with *place relations*. The selection included genres such as geography manuals and travel guides.

- Group 3: this group consisted of twenty texts characterized by a *narrative* discourse purpose and a subject matter dealing with *events* and *participants*. The texts included in this group were historical passages and biographies.

[2]The judges were two Ph.D. students and the author of this paper. Their contribution in the sample collection used for this study is acknowledged in Section 5.

- Group 4: this group consisted of twenty texts characterized by an *instructive* discourse purpose and a subject matter dealing with *steps in a procedure*. The texts included in this group were method sections from instruction manuals, and recipes from cookery books and magazines.
- Group 5: this group consisted of twenty texts characterized by a *persuasive* or *argumentative* discourse purpose and a subject matter dealing with *facts* and *ideas*. The genres included in this group were mainly essays and editorials.

Design and procedure

In order to explore the issues presented in the introductory section above, the corpus was analyzed as follows:

1. The different groups of texts were analyzed to specify the *chaining strategies* which organize them globally and locally, and their linguistic markers.[3] Different chaining strategies were operationalized in this study as follows:
   (a) *Characterization strategy*: line of development consisting of references to a concept or to characteristics of the generic class to which the concept belongs (including attributes and parts).
   (b) *Temporal strategy*: line of development consisting of temporal references which mark specific points along a temporal span.
   (c) *Participant strategy*: line of development consisting of references to participants in the text.
   (d) *Spatial strategy*: line of development consisting of references to locations along an imaginary tour through which the reader is guided, or to spatial points with respect to a central place of observation.
   (e) *Sequential strategy*: line of development consisting of steps in a procedure.
   (f) *Though- or counter-argument strategy*: line of development consisting of a series of arguments in favour or against the thesis defended by the writer.

   Each of these strategies was determined by the characteristic markers which define its development. Thus, for example, a temporal chaining strategy was determined by the temporal references which signal specific points along a

[3] The term 'chaining strategy' was originally proposed by Lavid [1994] to refer to characteristic lines of textual development which organize different types of texts. These lines of textual development have received different names in the literature: *text strategies* (Enkvist [1987a; 1987b]), *framing strategies* ([Witte and Cherry, 1986, 130ff]), *rhetorical designs* [Nash, 1980], *methods of development* [Fries, 1981], *orientations* of various kinds [Givon, 1984, p. 245], [Grimes, 1975], [Longacre, 1983, pp. 7–8], or *text-strategic continuities*, [Virtanen, 1992].

temporal span. In order to determine the chaining strategy which organized a given text, a search for characteristic references or scopes was carried out as follows:

- For the temporal chaining strategy, a search for temporal references signaling specific points along a temporal span was carried out.
- For the characterization strategy, a search for references to a concept or to characteristics of the generic class to which the concept belongs (including attributes and parts) was carried out.
- For the participant strategy, a search for references to different participants in the text was carried out.
- For the spatial chaining strategy, a search for spatial scopes or references to locations was carried out.
- For the sequential chaining strategy, a search for steps in a procedure was carried out.
- For the through-or counter-argument strategy, a search for arguments in a favour or against the thesis defended by the writer was carried out.

2. Each marker of a given chaining was counted in each text, and inspected to check whether it was thematized or not. If thematized, the semantic type of *theme* selected was annotated.
3. The data were statistically analyzed to determine whether there existed statistically significant correlations between the markers of a given chaining strategy and the semantic type of *theme* selected to signal those markers.

For illustration, Example 1 below will serve to show how the global spatial chaining strategy was determined:

EXAMPLE 1.

1. British Columbia is divided into four parallel ranges.
2. **To the east** the Rocky Mountains are steep and ragged with summit elevations sometimes exceeding 3,048 metres.
3. The mountains melt into the broad timbered Central Plateau region, an area marked by glacial movement.
4. The Coast Range, part of the North American Cascades, represents the largest mountain mass in Canada.
5. **Along the western coast** extends a protective chain of islands varying in size from Vancouver Island with its 32, 137 square kilometres (12,408 sq.m.) to smaller island, such as the Gulf Islands: Saltspring, Galiano, Mayne, Pender, Saturna, and Gabriola.

6. Almost all the islands are heavily wooded, forming a labyrinth of sheltered channels.

The text is characterized by a *descriptive* discourse purpose and a subject matter dealing with *place relations*. The global spatial strategy consists of two references to spatial points (spatial scopes) with respect to a central place of observation: the East and the West. These two references are encoded in the text by means of two locative expressions: *to the east, along the western coast* (boldface and underlined in the text), which act as markers of the chaining strategy. Note, that these two markers appear in thematic position as well. This is not always the case in all the texts analysed, where it is possible to find markers of a given strategy in non-thematic position. The thematization of the markers of a given strategy, is, therefore, a conscious option on the part of the writer which may or may not be chosen. The statistical analysis of the different text-type groups used in this study indicates, however, that, in general, certain markers are predominantly thematized to signal the strategy which characterizes a given text-type group.

3 RESULTS

The analysis of the *chaining strategies* which organize texts in the corpus showed that certain text type groups systematically select certain strategies to organize information. It also showed that certain strategies were used globally while others were used more locally, depending on the text types. This finding confirms Enkvist's claim according to which texts can exhibit multiple strategies [Enkvist, 1987a]. The results of the analysis of the global chaining strategies in each group of texts were the following:

1. In the first group of texts, i.e. those with an *expository* DP and dealing with generic concepts, the analysis showed that the characteristic global chaining strategy used by writers was the *characterization* strategy.
2. In the second group of texts, i.e. those with a *descriptive* DP and dealing with place relations, the analysis showed that the characteristic global chaining strategy used by writers was the *spatial* chaining strategy. Some texts were also steered locally through a *characterization* strategy.
3. In the third group of texts, i.e. those with a *narrative* DP and dealing with events and participants, two main global chaining strategies were typically used: the *temporal* chaining strategy, and the *characterization* strategy. The former predominates in historical passages, while the latter predominates in biographical passages. This is probably due to the emphasis either on the chronology of events in the former case, or on the life's episodes of the person, in the latter. A participant strategy was also present in some texts, but organizing information at a local level of analysis.

4. In the fourth group of texts, i.e. those with an *instructive* DP and dealing with steps in a procedure, the characteristic global chaining strategy used by writers was the *sequential* chaining strategy.

5. In the fifth group of texts, i.e. those with a persuasive or *argumentative* DP and dealing with facts and ideas, the characteristic global chaining strategy used by writers was the *through- or counter-argument* pattern. A *characterization* strategy was also found organizing information locally.

Table 1 below shows the distribution of global and local chaining strategies for each text type group:

Table 1. Text types and chaining strategies

Types	Texts	Chaining Strategies					
Groups	sample	Charac.	Spatial	Temporal	Part.	Sequential	Th.- C
1	20	20	0	0	0	0	0
2	20	10 (local)	20	0	0	0	0
3	20	5	0	15	5(local)	0	0
4	20	0	0	0	0	20	0
5	20	7(local)	0	0	0	0	20

Charac. = Characterisation; Part. = Participant; Th.- C. = Through- or Counterargument

As Table 1 illustrates, each text type group typically selects a specific chaining strategy to organize information globally, though it is possible to find other strategies working locally. Note that in the third group (i.e. those texts characterized by a *narrative* discourse purpose and dealing with *events* and *participants*), fifteen texts selected the *temporal chaining strategy*, while five opted for the *characterization strategy* to organize information globally, depending on the subject matter of the genres: in historical passages, the subject matter concentrated on the chronology events, and therefore, the temporal strategy was preferred; in biographies, the emphasis was more on the life of a person, and therefore, the characterization strategy was preferred.[4]

The second phase of the analysis consisted in inspecting the linguistic markers of the chaining strategy which characterizes a given text-type group, and analysing their correlations with the semantic types of *themes* selected to signal that strategy. This was done by counting each time a marker of a given strategy was introduced in the text analysed, and checking the type of *theme* selected to signal it. The results of the corpus analysis were the following for each group:

[4]The distribution of chaining strategies in Table 1 applies only to the text type groups analyzed in this study, which are characterized by specific combinations of the contextual variables presented before. It would be possible, however, to find a different distribution of chaining strategies if the combinarions of DP and SM were different. For example, if the discourse purpose is descriptive and the subject matter deals with objects, not with place relations, probably a characterization chaining strategy, rather than a spatial one, will be used.

1. In the first group of texts, i.e. those with an *expository* DP and dealing with generic concepts, *topical themes* were predominantly selected each time a reference was made to a concept or to attributes or parts of it. The Pearson correlation coefficient showed a statistically significant correlation between the number of referential scopes and the selection of *topical themes* ($r=0.947$; $p<0.05$), while no statistically significant correlation was found with any of the other theme types.

2. In the second group of texts, i.e. those with a *descriptive* DP and dealing with place relations, *locative themes* were predominantly selected each time a spatial reference was made in the text. The Pearson correlation coefficient showed a statistically significant correlation between the number of spatial scopes and the selection of *locative themes* ($r=0.979$; $p<0.05$), while no statistically significant correlation was found with any of the other theme types.

3. In the third group of texts, i.e. those with a *narrative* DP and dealing with events and participants, *temporal themes* were predominantly selected each time a temporal scope was opened in the text. When the chaining strategy was a *characterization* or a *participant* strategy, *topical themes* were preferred to signal those strategies. The Pearson correlation coefficient showed a statistically significant correlation between the number of temporal scopes and the selection of *temporal themes* ($r=0.970$; $p<0.05$), while no statistically significant correlation was found with any of the other theme types.

4. In the fourth group of texts, i.e. those with an instructive DP and dealing with steps in a procedure, both *process themes* and *temporal themes* (usually presented typographically by numbers) were predominantly selected each time a new step in a procedure was taken. Preconditions and goals were also selected as themes in instruction manuals. In recipes, only *process themes* are selected to signal the steps in the procedure, as these texts are experientially iconic and do not need to specify the sequence of processes temporally. The Pearson correlation coefficient showed a statistically significant correlation between the number of steps and the selection of *process themes* to mark that strategy ($r=0.953$; $p<0.05$), and also between the number of steps and the selection of *temporal themes* ($r=1$), while no statistically significant correlation was found with any of the other *theme* types.

5. In the fifth group of texts, i.e. those with a persuasive DP and dealing with facts and ideas, the types of themes selected were predominantly *text-builders* (i.e. conjunctives) each time a step in an argument was introduced in the text. The Pearson correlation coefficient showed a statistically significant correlation between the number of arguments and the selection of *text-builder themes* to mark that strategy ($r=0.610$; $p<0.05$), while no statistically significant correlation was found with any of the other theme types.

4 DISCUSSION

As the results of the empirical analysis have shown, certain text types, characterized by a specific combinations of discourse purpose and subject matter, present a maximal probability to exploit a global chaining strategy to organize their information. For example, texts characterized by a *descriptive* discourse purpose and dealing with place relations are typically organized globally by means of *spatial* chaining strategy, while texts with an *expository* discourse purpose and dealing with whole classes of objects or generic concepts tend to be globally organized by means of a *characterization* strategy. This does not preclude the use of other strategies working at a more local level of analysis, or when the combinations of discourse purpose and subject matter are different from the ones investigated in this paper.

Another important finding is the existence of statistically significant correlations between the global chaining strategies which characterize specific text types, and the semantic type of *themes* selected to signal those strategies. For example, texts which select a *temporal* chaining strategy to organize information globally tend to signal that strategy by means of *temporal themes*, while texts globally structured by means of a spatial strategy tend to signal it by means of *locative themes*.

These results suggest that the contextual variables of discourse purpose and subject matter investigated in this paper are important contextual constraints for the overall textual organization of information in discourse in general, and for its thematic patterning, in particular. The statistically significant correlations between global chaining strategies and thematic selection also provide empirical evidence for the claim that thematization fulfills a signaling role in discourse of the chaining strategy selected by text producers to organize textual information.

These findings are of both theoretical and applied interest. From the theoretical point of view, this empirical study makes considerable progress in exploring the relationship between specific contextual variables and the phenomenon of thematization in discourse, thus partly overcoming the limitations of previous studies which explored related issues (see [Fries, 1995, p. 13]).

From the applied point of view, the empirical findings can be fruitfully used in the application context of automated text generation, where the specification of contextual sources of control over textual variants is a fundamental task for the production of pragmatically-motivated texts. Multistratal generation architectures, such as the Komet–Penman text generation system (Teich et al., forthcoming), which can recognize the complexity of linguistic resources necessary for the generation of a wide variety of texts, can make use of these empirical findings to control the generation of different textual variants. These generators can use various representational means to specify empirical results. These are:

- The *system network*: this is a feature hierarchy consisting of linguistic types. Though typically used in Penman-style generators to represent grammatical information, it has also proven useful at the text-planning level to represent discourse information [Hovy *et al.*, 1993]. In recent work, the system network is also used to specify an interface which mediates between the account of grammatical Theme -as the one found in the very large computational systemic-functional grammar of English described by C. Matthiessen [1995]—and its discourse semantics by creating a semantic system for *theme* in English which recognizes the types of information which the discourse resources will preselect [Ramm *et al.*, 1995; Lavid, 1998; Lavid *et al.*, 1999]. Such a system will function as an interlevel between lexicogrammar and discourse semantics abstracting from grammar and realizing text-building categories.

- *The mechanism of preselection.* This can be used to express type constraints associated with the types in the system network across strata. For example, if at the contextual level we have the information that the DP is narrative and that the SM deals with events, the chaining strategy (represented at the discourse level) can be constrained to be of the temporal type, and the *theme* type (represented at the semantic level) to be of the temporal type as well.
- *The inquiry framework.* This mechanism, developed by Mann and others [1983], and explained elsewhere [Matthiessen and Bateman, 1991], has been used at the interface with the grammar to specify interstratal constraints. Given its flexibility, it can also be used to represent as inquiries the contextual motivations under which particular options are selected from a given system network [Lavid, 1998].

Clearly, it is not the aim of this paper to provide a detailed specification of how the results of the empirical analysis could be implemented. It can only be emphasized here how crucial it is for any text generation system which aims at generating pragmatically-adequate texts to recognize and represent the results of empirical investigations such as the ones presented in this paper. A step towards that goal could be the adoption of a multistratal generation architecture where responsibilities are distributed across different levels of description.[5]

Interesting directions to extend the results of this initial investigation would be a cross-linguistic analysis of how the relationship between contextual variables and thematization works in other languages, or how other contextual variables -different from the ones studied here- are related with the phenomenon of thematization. While the former is currently being explored [Lavid, 1998; Lavid *et al.*, 1999], the latter is a matter for future research.

ACKNOWLEDGEMENTS

I would like to thank the Ph.D. students Isabel Alonso and Francisco Ballesteros for their contribution in the collection and analysis of part of the texts used for

[5]For a descriptive account of such an attempt see [Lavid *et al.*, 1999].

this study. The work presented in this paper is an extension and refinement of a preliminary study carried out by the author [Lavid, 1994] within the Esprit BR Project 6665 Dandelion (Discourse Functions and Discourse Representation: An Empirically and Linguistically Motivated Interdisciplinary Approach to Natural Language Texts).

Universidad Complutense de Madrid, Spain.

REFERENCES

[Bäcklund, 1990] I. Bäcklund. Theme in English telephone conversation. Paper delivered at the *17th International Systemic Congress*, Stirling, Scotland, June 1990.

[Berry, 1989] M. Berry. Thematic options and success in writing. In *Language and Literature-Theory and Practice: A Tribute to Walter Grauberg*. C. Butler, R. and J. Cardwell, eds. University of Nottingham, 1989.

[Enkvist, 1987a] N. E. Enkvist. Text strategies: single, dual, multiple. In *Language topics: essays in honour of Michael Halliday*. Vol.II. R. Steele and Threadgold, eds. John Benjamins, Amsterdam, 1987.

[Enkvist, 1987b] N. E. Enkvist. More about text strategies. In *Perspectives on Language Performance. Studies in Linguistics, Literary Criticism and Language Teaching and Learning*, W. Lörscher and R. Schulze, eds. Günter Narr, Tübingen, 1987.

[Fairclough, 1989] N. Fairclough. *Language and Power*. Longman, London, 1989.

[Firbas, 1992] J. Firbas. On some basic problems of Functional Sentence Perspective. In *Advances in Systemic Linguistics: Recent Theory and Practice*. M. Davies and L. Ravelli, eds. Pinter, London, 1992.

[Francis, 1989] G. Francis. Thematic selection and distribution in written discourse, *Word*, **40**, 201–221, 1989.

[Francis, 1990] G. Francis. Theme in the daily press. *Occasional Papers in Systemic Linguistics*, **4**, 51–87, 1990.

[Fries, 1981] P. H. Fries. On the status of theme in English: arguments from discourse. In *Micro and Macro Connexity of Texts*. Petöfi and Sozer, eds. Buske Verlag, Hamburg, 1981.

[Fries, 1995] P. H. Fries. A personal view of theme. In *Thematic development in texts*, M. Ghadessy, ed. pp. 1–19, Pinter, London, 1995.

[Ghadessy, 1995] M. Ghadessy. Thematic development and its relationship to register and genres. In *Thematic Development in Texts*, M. Ghadessy, ed. pp. 129–146. Pinter, London, 1995.

[Givon, 1984] T. Givon. *Syntax: A Functional-typological Introduction*. Volume 1. John Benjamins, Amsterdam, 1984.

[Grimes, 1975] J. E. Grimes. *The Thread of Discourse*. Janua Linguarum, Series Minor 207. Mouton, The Hague and Paris, 1975.

[Halliday, 1967] M. A. K. Halliday. Notes on transitivity and theme in English, *Journal of Linguistics*, **3**, 37–81, 1967

[Halliday, 1978] M. A. K. Halliday. *Language as Social Semiotic: The Social Interpretation of Meaning*. Edward Arnold, London, 1978

[Halliday, 1985] M. A. K. Halliday. *An Introduction to Functional Grammar*. Edward Arnold, London, 1985.

[Hovy *et al.*, 1993] E. Hovy, J. Lavid, V. Mittal and C. Paris. Employing knowledge resources in a new text planner architecture. In *Aspects of Automated Natural Language Generation*, R. Dale, E.Hovy, D. Röstner, and O. Stock, eds. pp. 57–72. Springer, 1993.

[Hymes, 1974] D. Hymes. *Foundations in Sociolinguistics: an Ethnographic Approach*. University of Pennsylvania Press, Philadelphia, PA, 1974.

[Kress and Threadgold, 1988] G. Kress and T. Threadgold. Towards a social theory of genre. *Southern Review*, **21**, 215–243, 1988.

[Lavid, 1994] J. Lavid. Theme, Discourse Topic and Information Structuring. Deliverable R1.2.2b of WP1.2.2. Esprit Basic Research Project 6665 DANDELION, Universidad Complutense de Madrid, Madrid, October 1994.

[Lavid, 1998] J. Lavid. The relevance of corpus-based research for contrastive linguistic and computational studies: thematization as an example. In *Corpora in Semantic and Pragmatic Research*. pp. 117–139. Publicaciones del Instituto de Lingüística Aplicada. Barcelona: Universidad Pompeu Fabra, 1998.

[Lavid *et al.*, 1999] J. Lavid, W. Ramm and C. Villiger. Thematization in English and in German: empirical and computational findings. Submitted for publication.

[Leckie-Tarry, 1995] H. Leckie-Tarry. *Language and Context: A Functional Linguistic Theory of Register*. Pinter, London and New York, 1995.

[Lemke, 1988] J. L. Lemke. Text structure and Text Semantics. In *Pragmatics, Discourse and Texts: Some Systemically-inspired Approaches*, Steiner and Veltman, eds. Pinter, London, 1988.

[Longacre, 1983] R. E. Longacre. *The Grammar of Discourse*. Plenum Press, New York and London, 1983.

[Mann, 1983] W. C. Mann. An overview of the the Penman text generation system. Technical Report SI/ RR-83-114, USC/Information Sciences Institute, Marina del Rey, CA, 1983.

[Martin, 1995] J. R. Martin. More than what the message is about: English Theme. In *Thematic Development in Texts*, M. Ghadessy, ed. pp. 23–258. Pinter, London, 1995.

[Matthiessen, 1995] C. M. I. M. Matthiessen. *Lexicogrammatical Cartography: English Systems*. International Language Sciences Publishers, Tokyo, 1995.

[Matthiessen and Bateman, 1991] C. M. I. M. Matthiessen and J. Bateman. *Text Generation and Systemic-Functional Linguistics: Experiences from English and Japanese*, Pinter, London, 1991.

[Nash, 1980] W. Nash. *Designs in Prose*. Longman, London, 1980.

[Ramm *et al.*, 1995] W. Ramm, A. Rothkegel, E. Steiner and C. Villiger. Discourse Grammar for German. Deliverable R2.3.2 of WP 2 *Grammar Integration*, ESPRIT Basic Research Project 6665 DANDELION, University of Saarland, Saarbrcken, 1995

[Rubin, 1984] D. L. Rubin. The influence of communicative context on stylistic variation in writing. In *The Development of Oral and Written Language in Social Contexts*, Pelligrini and Yawkey, eds. Ablex, Norwood, NJ, 1984.

[Teich *et al.*, forthcoming] E. Teich, J. Bateman and L. Degand. (forthcoming). Multilingual textuality: experiences from multilingual text generation. In *Selected Papers from the Fourth European Workshop on Natural Language Generation*, Pisa, Italy, M. Zock and G. Adorni, eds.. Springer, Berlin, forthcoming.

[Virtanen, 1992] T. Virtanen. *Discourse Functions of Adverbial Placement in English*. Abo Akademi University Press, 1992

[Werlich, 1983] E. Werlich. *A Text Grammar of English*. Quelle and Meyer, Heidelberg, 1983.

[Wang, 1991] L. Wang. Analysis of thematic variations in Buried Child. Paper delivered at the *First Biennial Conference on Discourse*. Hangzhou Peoples Republic of China, June 1991.

[Witte and Cherry, 1986] S. P. Witte and R. D. Cherry. Writing processes and written products in composition research. In *Studying Writing: Linguistic Approaches. Written Communication Manual 1*, C. R. Cooper and S. Greenbaum, eds. Sage, Beverly Hills, 1986.

MARK GALLIKER AND DANIEL WEIMER

CONTEXT AND IMPLICITNESS: CONSEQUENCES FOR TRADITIONAL AND COMPUTER-ASSISTED TEXT ANALYSIS

1 INTRODUCTION

The present study discusses the concept of context and implicitness using the example of verbal discriminations in private and public discourses. The investigation begins by distinguishing between explicit and implicit meaning. The central question of our research is whether explicit indicators for implicit meanings exist, and whether the relationship between explicitness and implicitness has methodological relevance. The ultimate goal of the investigation is to discover some clues which can assist us in pursuing qualitative and quantitative context analysis.

One of the starting points of the study is the ordinary-language approach. Words obtain meaning in relation to their social and/or verbal context, functioning as more than merely references to things or ideas. This phenomenon is often so obvious that one does not pay particular attention to it (cf. e.g. [Wittgenstein, 1971; Galliker, 1977; de Certeau, 1988]). However, natural speech is understood not as static, but as dynamic in the sense of discourse analysis in the study at hand (cf. e.g. [van Dijk, 1985; Jäger, 1993; Luutz, 1994]). Special attention is given to a dialectical moment in the relationship between text (or speech) and context. The latter is understood as an external complement of the internal meaning that is perpetually reconstituted (cf. e.g. [Markov'a, 1990]). Verbal units (words, sentences) do not have fixed meanings, but change their meanings depending on which other verbal units appear in the context. Therefore the primary consideration in the interpretation of this order is its relation to other units or significations. The derivation of and grounds for this primacy of the signification will not be discussed here (cf. [Galliker, 1990]). We will not, however, neglect the important technical issue of computer readability of the significations, since this issue is especially relevant for the quantitative aspects of context analysis.

2 CONTEXT

The context of a certain utterance is the (cultural and social) surroundings or circumstances in which the utterance occurs (cf. e.g. [Wittgenstein, 1980]). The context has artificial character because it is only relevant as meaning is developped, organized, produced, and/or reproduced, by human beings. Thus the circumstances can also be understood as a framework which embodies human meaning. (cf. e.g. [Lang, 1992]). Conversely, language use is very largely prescribed

P. Bonzon, M. Cavalcanti and R. Nossum (eds.), Formal Aspects of Context, 49–63.

and proscribed by the situation in which it is spoken or written, which can be understood as the 'language reflects context' paradigm (cf. e.g. [Gumperz, 1992; Giles and Coupland, 1991]).

According to the point of view of the subject, the context is always restricted to some degree. If the restriction is eased the context can expand infinitely. The salient context can be the text in direct relationship to the utterance in question, i.e. the text as a whole in which the utterance appears, and in particular the text immediately before the utterance. The context cannot, however, be reduced to this alone. There are different contextual levels, ranging from the culture as a whole to the face-to-face interaction of human beings [Bronfenbrenner, 1979; Wagner, 1997]. The content of the same utterance focused in different contexts obtains different meanings. If the participants do not agree on the same level (or focus on the same episodes or contextual sections), communication problems can arise (cf. e.g. [Forgas, 1988]). Different possibilities for interpretation often result in lengthy discussions. Contradictory interpretations which appear consecutively in alternation can be termed as *ambiguous* meanings [Galliker and Wagner, 1995].

3 THE EXAMPLE OF VERBAL DISCRIMINATION

Social discrimination can be defined as unequal treatment and valuation of persons according to the cultural and social categories to which they belong or are perceived to belong. *Categorization* means using a generalized category (for example *'asylum-seekers'*) when referring to an individual. The individual is no longer regarded as a person with various interests, behavioural and psychological characteristics, but is 'frozen' into a prototype. *Valuation* means the assignment of a label according to the societal acceptance of the group in question. In the case of negative valuation, a pejorative designation (*'sham asylum-seekers'*) or a pejorative judgement (*'They are only after our money'*) is made. Unlike an insult, the devaluation directed against a person by categorization refers not only to that person's character or conduct but also to his/her cultural and social identity.

Social discrimination expressed in private or public discourse is called *verbal discrimination* [Wagner *et al.*, 1993]. Unlike verbal discrimination in face-to-face discourse, in public discourse there is a distance between discriminator and victim. Furthermore there is no control over the identity of the recipients [Graumann, 1995]. Verbal discrimination is constituted in a multifaceted manner [Graumann and Wintermantel, 1989; Galliker *et al.*, 1994], consisting essentially of categorization and evaluation. These claims are based on the results of two validation studies, one involving 54 German and 54 immigrant subjects [Galliker *et al.*, 1994], and the other involving 58 German subjects [Galliker *et al.*, 1995].

Verbal discrimination can be either explicit or implicit. *Explicit discriminations* are utterances expressing judgments of 'outgroup'-categories. Such a judgment is made lexically, for example:

'Asylum-seekers are lazy', or by the content of the utterance as a whole, for ex-

ample: *'There are no hard-working immigrants'*. Here, the judgment is negative, but it can also be a negation of a negative judgement, e.g. *'These asylum-seekers are not lazy'*, or a positive judgment, e.g. *'These immigrants are hard-working people'*. *Implicit discriminations* are utterances in which the judgment, and perhaps other discriminatory aspects, are not reflected in the words or the content of the utterance. They can only be understood as discriminatory from the context in which they are made [Galliker *et al.*, 1995; Wagner, 1997]. The utterance *'They are very busy people'* may be a positive judgment, if the context suggests social norms in which work is appreciated, but it may be a negative judgment if, in a particular social situation or subgroup, the conventional norm of work is not appreciated [Galliker and Wagner, 1995].

4 THE DISAPPEARANCE OF VERBAL EVALUATION

Implicit discrimination is characterized by the disappearance of verbal evaluation. The following examples illustrate this development from explicit to implicit speech:

(1) *'Refugees are lazy'*.

(2) *'Refugees are not hard-working people'*.

(3) *'Apparently we have to pay for more than our share'*.
(If said, for example, in a discussion about refugees.)

(4) *'We have to work more'*.
(As above.)

In (1), the devaluation appears as an expression 'lazy', while in (2) the devaluation is the phrase 'not hard-working people'. Both (1) and (2) are obviously negative evaluations and thus explicit discriminations, but one finds the pejorative term 'lazy' only in (1). This illustrates the differentiation between *expression* and *content* according to the logician Frege [1892/1994]. The utterances (3) and (4) are not discriminatory utterances per se, because they are devoid of an explicit devaluation. When put in an appropriate context, however, they can become quite discriminatory. The devaluation of refugees who cannot provide for themselves emerges from the relation between the utterance and its verbal context. The devaluation results from the process of inference [Graumann, 1995]. In (3), the word 'apparently' refers to the context in which that utterance is made. In this case the speaker verbally signals that the context is important (cf. [Frege, 1892/1994]). This kind of verbal cue is not present in (4). Nevertheless, the devaluation can be deduced from the context of the discourse. The question is whether a listener can make such a deduction in the flow of ordinary discourse. The speaker in (4) may not signal the context by verbal cues, although he/she may instead opt to use paralinguistic and non-verbal cues when referring to context. Cues are assigned a value by the tacit awareness of their socially shared meanings; this, in turn, refers to contextual conventions [Gumperz, 1992].

In today's public discourse, open devaluation of 'outgroups' is viewed in a very negative way. Van Dijk [1991] points out that even in so-called informally racist countries certain standards of tolerance and non-discrimination apply to public speech. Pettigrew [1989] focuses on modern forms of prejudice formation, particularly in the USA. He distinguishes between a subtle language of prejudice, a blatant language of prejudice, and a prejudice-free language. Blatant forms of verbal discrimination contradict accepted social norms, while subtle forms of discrimination do not infringe on these. In public discourse explicit discrimination against minority groups is suppressed according to norms of *political correctness*. An intense political discourse on verbal discrimination and political correctness has evolved in the past few years. Hughes [1995] claims that political correctness leads to euphemistic discourse without any change in social life. While problems of symbolic racism may be resolved [Dovidio and Garetner, 1985], the problems of the producers and recipients of verbal discriminations remain.

One can assume that political correctness can bring about increased implicit discrimination in public discourse. This assumption was confirmed in several studies that focused on the analysis of newspaper reports [Galliker *et al.*, 1998; van Dijk, 1994], and parliamentary debates [Galliker and Wagner, 1995]. The substitution of explicit discrimination by implicit discrimination cannot necessarily be seen as an indication of decreasing prejudice and racism in society. However, it may be a sign that social discrimination is commonplace. In this case social discrimination would be nearly omnipresent without being considered or labelled discrimination, and thus not verbalized in discourse. Implicit discriminations rely on mutual feelings and shared cognitions. Since they are triggered by verbal or non-verbal cues, they operate without having been articulated.

Implicit discriminations do not seem to be less politically effective than explicit ones. In fact, the reverse seems to be true, since they are more difficult to perceive, check, and identify. In today's society, open discrimination seems to have a direct effect primarily on people with an extremely right-wing ideology. Some studies indicate that, in the long run, more subtle forms of discrimination are very effective in a broader context [Pettigrew and Meertens, 1995].

5 SOCIAL KNOWLEDGE

Often the salient context is not a concrete situation, as in the example above. An understanding is nevertheless possible, since speaker and listener normally share a certain amount of cultural and social knowledge. The importance of rhetorical features in social representations was examined by Billig [1988; 1993]. Shared representations are the basis of every communication, forming an important condition for implicit communication [Wagner, 1997].

If the speaker believes that the listener shares some knowledge, he/she may verbalize only the information that is necessary for the interpretation of the message. Commonplace assumptions or beliefs, results of discourses with the same or similar participants, and what is regulated in the utterances of the actual discourse all contribute to the additional knowledge necessary for interpreting the utterance in question. If in doubt, the recipient may look for cues referring to the supposed knowledge. Consider the following example as clarification: an immigrant German from Romania reports, *'We got along with everybody, even with Jews'*. The utterance is spoken with pride and in a positive sense. However, the word 'even' indicates that what is being said is surprising (cf. [Sinclair, 1992]). The surprise is based on the comparison with the common understanding. Apparently this common sense comprises the agreement that one can not get along with the Jews as one would with members of other religions, or with other people in general. This means that the significance of the main sentence ('We got along with everybody') is enhanced by mentioning Jews. The context assumed here implies that contact made with members of that group may generally be unpleasant. Thus the ostensibly friendly statement is, in fact, anti-Semitic.

The verbal cues are, in most cases, particles or modal words (e.g. so, even, naturally, certainly). German handbooks [Helbig, 1990; Helbig and Helbig, 1993] and dictionaries of English usage [Fowler, 1983; Sinclair, 1992] give some information about the cognitive and emotional functions of such (seemingly) meaningless words. In particular, there is a description of which kind of context or implied knowledge is indicated by a special particle or modal word.

If the speaker uses verbal cues it is less difficult to identify implicit discrimination, because these cues indicate that the content of the entire communication package is more than the verbal communication. The listener is thus advised by the speaker to look for other meanings regarding the context. Without verbal cues (and especially in the extreme case when there are no non-verbal cues available; see above), no indication exists that the utterance contains an additional meaning. Yet it remains uncertain whether, or to what extent, the listener may be influenced by the implicit meaning. In such cases, it may be an unintentional influence, and/or without recipient-reflection [Galliker and Wagner, 1995].

In a certain political context the mere mentioning of the 'outgroup'-category, e.g. *'Jews'*, seems to be enough to renew verbal discrimination against the group. In this connection, one would not recognize the devaluation because it is inherently clear, and it would be 'natural' that members of the 'outgroup' are not treated in the same way as members of the 'ingroup'. That is not labelled as discrimination for most people, since it first appears neutral and not as a negatively charged term which differentiates between members of the 'outgroup' and human beings.

Shared cultural and social evaluations are very important in public discourse. The speaker usually presumes, without saying or thinking, that the listener possesses evaluative dimensions similar to his/her own. The speaker's own negative and/or positive judgments are based on these dimensions. Listeners with similar dimensions are capable of understanding these judgments, regardless of their individual opinions. If necessary, they can even interpret a matter-of-fact representation, thereby separating the 'ingroup' and 'outgroup' in terms of negative or positive discrimination.

6 EXTRACTION OF THE IMPLICIT MEANING

For a certain period of time verbal cues (and/or non-verbal ones not discussed here) may be sufficient to confirm common evaluative dimensions. Even putting matter-of-fact remarks simply, without any cues, may sometimes suffice in a particular context. In such cases there are no 'local traces' of implicit discriminations. There is nothing that would advise the listener to pay attention to something that is embedded in the interaction but cannot be heard directly. Although such tacit emotions are not expressed in discourse; they can continue in an undifferentiated state. This may be one of the reasons many people were surprised at the sudden resurgence of violence against 'outgroups' following German reunification [Galliker and Wagner, 1995; Galliker *et al.*, 1996].

Certainly, implicit discriminations cannot exist completely without explicit discriminations. Although the speaker normally supposes that the listener shares his/her evaluative dimensions, this is frequently not the case. This can be explained by cultural and social differences. The evaluative dimensions of others can be the reason for uncertainty in interpersonal relations. If this uncertainty is not reduced, communication will lack coordination and tend towards fragmentation [Berger and Bradac, 1982].

How can this be avoided? It is necessary to confirm the supposed evaluations in discourse from time to time by stating them explicitly, such as after an ambiguous situation. Additional explicit utterances, or at least additional verbal cues indicating the salient context, are the basis for confirming a mutual agreement. They are also often necessary for continuing a dialogue.

Explicit discriminations, corroborating implicit ones, seem to occur frequently as negations of devaluations or as positive evaluations. The remark *'These asylum seekers are not lazy'* assumes that in general, asylum seekers are lazy. The remark *'We have honest and very clean asylum-seekers'* sounds like an attempt to enhance the asylum-seekers' status. However, one can also interpret this remark as basically devaluating the entire refugee group. Positive discriminatory remarks allow for confirmations of dimensions in which 'outgroups' are generally evaluated as worse than 'ingroups'. This is why they perpetuate prejudice and implicit discriminations. At the same time the speaker may believe that he/she is acting in a humane way. The listener rarely questions such remarks because it would be difficult to establish proof of verbal discrimination. One can also use a similar strategy by devaluating members of the *'ingroup'*; for example: *'These young Germans are not as hard-working as the foreign apprentices'*. Placed in the proper context, this remark could imply that, as a rule, young Germans work harder than young migrant workers in the Federal Republic.

Evaluations of social groups often tend to emerge through particular constellations of significations in the context of relevant categories. In these cases particular representations and expressions are introduced into the discourse at a later stage. When these extractions are more or less socially shared, constant replication is no longer necessary. From time to time implicit discriminations are indicated by verbal cues or confirmed by explicit discriminations. Implicit discrimination is characterized by the circumstances under which it takes place, rather than by the so-called inner experiences of the speakers and listeners. An accurate study of the given cues and explications is necessary in order to discover an implicit meaning of the discourse.

7 TEXT EXAMPLE

The relationship between implicit and explicit discrimination can be illustrated using an article written by Paul Johnson which appeared in the British newspaper *Daily Mail*, on July 1, 1995:

> *"Is Western Man Becoming an Endangered Species?"* (pp. 8 ff.).

Beginning with the title, the author produces a context which highlights the protection of one category of people. Focusing on the 'ingroup', an open devaluation of the 'outgroup' cannot be identified. The category for the 'outgroup' is introduced only in the 13th sentence by the assumption *'that the percentage of white Britons will fall, and those of blacks and Asians will rise'*. The categorization appears in this phrase without explicit devaluation of the 'outgroup'. Nevertheless, keeping to the theme of 'endangered species', the rising population of the 'outgroup' must appear as something akin to a threat. This interpretation is confirmed by the following sentence: *'That strikes apprehension into many, perhaps most, people'*. In this expression of the threat nothing appears which is clearly against people from other parts of the world. The 'outgroup' has a high birth-rate, which is evaluated positively by the writer; the fact that the 'ingroup' has a low birth-rate is the problem. There is no explicit devaluation, but the negative evaluation of the 'outgroup' in relation to the 'ingroup' seems to be implied. Later in the text, the expansion of European people in past centuries is described:

> *'From the 11th century, thousands of ambitious young men from fertile France roamed the Mediterranean looking for land. ... That venture failed in the end, but from the 15th century onwards, the Portuguese and Spanish expanded first into the Atlantic Islands such as Madeira and the Canaries, then crossed the ocean to the Americas. They were soon followed by the French and the British, and eventually by the Germans and Italians. Europe exported colonists for 500 years, overwhelming or exterminating the native inhabitants.'*

The last sentence of this passage can be interpreted as a negative discrimination against the *'ingroup'*. The reason for colonialism is not simply 'Europe', nor *all countries* of Europe, and of course not *all people* of Europe, but rather the accumulation and concentration of money and therefore power, making it possible to organize and carry out trans-Atlantic slavery. This was all orchestrated by increasingly powerful states in Europe as well as Africa [Walvin, 1992]. The next sentence focuses on the results of the colonialism:

> *'North, South and Central America as they exist now are virtually European creations.'*.

The word 'creation' has a positive connotation. Thus, the 'ingroup' ('European') is evaluated positively at this point. In the following sentences the writer reverts to the evaluation of the 'outgroup', as was formerly implied in the text:

> *'Now all these trends have been set into reverse. Europeans still emigrate in limited numbers. But they no longer colonize. Indeed, nearly all the colonies are now independent and it is the Africans, Asians and even Latin Americans who are coming to Europe.'*

The last sentence of this passage is an implicit discrimination. In light of the previous text the sentence seems to acquire a negative discriminatory meaning. The context is established in the first three sentences of the passage and in the first phrase of the sentence in question. The first word of the first phrase ('indeed') emphasizes what came beforehand, indicating how things are according to the common understanding. The second phrase of the sentence ('it is the Africans, Asians and *even* Latin Americans who are coming to Europe') achieves an implicit devaluative meaning, because the arrival of the people from other continents is treated as equivalent to the colonialism described in the text above, meaning in the verbal context of the above passage. The particle 'even' indicates an exceeded expectation, an expectation that hasn't been exceeded yet or supposition reproduced in the text previously. Those seeking asylum come from countries far away from Europe, just as the Europeans previously invaded fa raway lands. It appears that the treatment of the 'outgroup' equals the treatment of the 'ingroup'. Nevertheless, the 'outgroup' is not described in particularly negative terms up to this point. The next sentence is:

> *'When Italy and Spain had high birth rates in the Thirties, they also had a high levels of emigration. Many of the emigrants were 'planted' in Libya. Now Italy and spain are targeted by young Muslims from Africa and Middle East as prime areas of settlement'*.

The 'outgroup' in this passage is not expressly devaluated. On the contrary: in regards to the dimensions of birth rate and settlement, the 'outgroup' is evaluated more positively while the 'ingroup' is evaluated negatively. The connection between settlement and colonialism (see above) is not reflected. An example for migration follows:

'Several times in recent years, for instance, migrants from Albania, which is technically in Europe but has an Afro-Asian-style high birth-rate, have tried to carry out mass-landings on the coast of Southern Italy'.

It seems that the writer can no longer categorize and evaluate in a clear manner. The Albanians belong on the one hand to the 'ingroup', and on the other hand to the 'outgroup'. The evaluation becomes ambiguous: *'high birth rate'* has more of a positive connotation; whereas *'mass-landings'* has stronger negative connotations for the author and his readers. It is exactly at this point that the usually implicit devaluation of the 'outgroup' is 'confirmed' by an explicit discrimination:

'Across the Mediterranean and the Balkans, then running deep into Central Asia, is an Arc of Aggression, as fertile peoples, nearly all of them Muslim, try to push into low-birth rate Christendom—by legal, and illegal migration, and sometimes by force'.

The implicit devaluations in the above sentences are extracted (*'Arc of Aggression'*). The interpretation of this as discriminatory phrases can only be substantiated by a single explicit discrimination, although the article as a whole makes the impression of being xenophobic. As already pointed out, explicit discriminations are rare in public discourse, but analogous effects can be attained by implicit discriminations, even when these effects are less noticeable (see above).

8 CONCLUSIONS

In summary, the possibilities of verbal discrimination can be formulated as follows:

Verbal discrimination =
[in-/outgroup] categorization +
[(negated/confirmed) positive/negative] evaluation.

Categorization and *devaluation* are of primary significance to discrimination. Whether the 'ingroup' or the 'outgroup' is positively or negatively evaluated is of secondary relevance as is the question of whether or not the corresponding judgments are negated or confirmed. In each case the identical structure of categorization and evaluation (i.e. the identical evaluative dimension) is of fundamental importance.

Furthermore, categorization and evaluation can be more or less closely linked to one another. One can only speak of an explicit discrimination with direct word sequence (*'Criminal foreigners should be deported'*) or when there is a direct association, such as one found within the same sentence (*'Foreigners are criminal'*). If the evaluation appears only in the neighboring context of the sentence containing the categorization (e.g. in written language in the same paragraph or the same text), then we are witnessing an implicit discrimination (see more in [Galliker, 1996]).

What is the significance of this for methodology? To identify the latent meaning of a verbal unit (e.g. a sentence) about minorities, one must view it in context with other parts of the discourse, although these parts may not be viewed together by the speaker or writer. In order to find a plausible interpretation one has to search for identical or similar sentences as well as for differing or even opposite sentences within the same dimension. It may be necessary to locate these at completely different places in the text. For example one can find sentences about migrants in a text corpus which are positive, negative or non-evaluating with regard to one another. As a first step positive and negative evaluations can be compared with one another qualitatively. With regard to identical or similar sentences this can also be done quantitatively.

Matter-of-fact sentences often acquire an evaluative meaning in the context of other sentences. Moreover, one can examine whether these implicit meanings are consistent, or inconsistent, with explicit meanings, and also whether there are sentences with a transitory function. As already illustrated for print-media discourse, in the case of implicit meanings the 'outgroups' are often negatively evaluated. The explicit meanings are either negations of devaluations (*'Many foreigners are not criminal'*), positive judgments of the 'outgroup' (*'There are even [!] foreigners that are very active in the fight against crime'*), or negative judgments of the 'ingroup' (*'There are German criminals too'*). The use of such sentences resembles a ritual devised to keep away 'evil spirits', and which may lead to a resurgence of aggressive tendencies.

Every sentence appearing in a context consisting of more than matter-of-fact expressions can, in principle, become an implicit discrimination. It is a type of metaphoric process that can occur spontaneously, or on purpose. Via this process, the categorizations are more or less positively or negatively placed, which means that they are evaluated. The relationship between the context and the sentence in question will appear sooner or later in a particular sentence. It is a kind of manifestation of that which can be understood as being *within* the discourse. The dimension of evaluation remains the same, but the manifest meaning can be the reverse of the implicit one. The extraction of the tacit presuppositions of verbal occurrences can also have the same meaning as the presuppositions. Both possibilities could be thoroughly documented with the qualitative analysis of 102 cassette recordings of spoken language. Sixty to ninety minutes of confidential, client-centered interviewers were recorded with clients who came from twenty different vocations including Bank presidents, labor uniopn leaders, tax officials, police officers, specialits and assistants [Galliker, 1980].

The manifestations of implicit meanings frequently do not appear in one of the following sentences, but at other places in discourse. The negative or positive evaluation must not necessarily refer to the 'outgroup', but can be shifted to the 'ingroup' or to other groups or persons, and occasionally to objects that do not appear to have even the remotest connection to the original reference. Shifted or displaced explications contribute more or less to new contexts which are relevant for new implicit meanings. So seems to work the natural language process.

The distance to the original categories can be determined through the use of various text and counting units (e.g. sentences, paragraphs, articles). In this way that which appears in the statement also appears in the context of this statement, and vice versa. This basically allows for two conclusions concerning the approach sketched here: (1) There is no absolute distinction between text and context in certain ways comparable to the painting of Paul Cezanne. (2) The distribution of words can be ascribed to the shifts, for example through dissociation and concentration. This makes it more apparent that it will indeed be possible to empirically investigate the so-called unconscious.

Similar to the grounded theory (e.g. Strauss, 1991), the approach already mentioned begins with *materials* as its starting point, in our case words, sentences and other language units that are produced and used by writers or speakers. In contrast to the grounded theory, however, the concepts of the approach sketched here are not only constructed as close to the foundation as possible (in order be nothing other than an external application on this particular foundation), which is then tested on new, often additionally obtained survey material (theoretical sampling). One allows instead the material to 'speak for itself' as much as possible in that one measures the verbal producfions of the same people (or groups) and texts against each other, allowing concepts to be successively explicated as far as necessary just as the content portrays a perpetual explication.

One can establish whether through constellations of existing significations other meanings are suggested, which become obvious only at another point in the text. Thus the way in which the verbal productions begin to dissociate and possibly contradict one another is easier to observe. What is understood as theory in this approach should, through the process of analyzing human productions, help make conscious the assumptions which are inherent in these productions, and which were initially still dormant. The theory genesis therefore does not rely on the usual distinction between discovery and substantiation relationships and is understood more as an extraction than as a mere reduction of empiricism.

The above discrimination formula can also serve as the basis for more quantitatively oriented studies. Since for several years the most important daily newspapers have saved practically their entire production on CD-ROM and made it publicly available in this form, *full-text researches* for verbal discriminations in public discourse are now possible. The first step is to determine the words relevant to the person categorizations and evaluations on the basis of the chosen approach and the establishment of word frequencies. With the help of computer programs such as TEXTPACK [Züll *et al.*, 1991] one can calculate *co-occurrences*, i.e. the appearance of two or more verbal occurrences (word categories) within a single text unit. These counting units of co-occurrence-analysis are word pairs, sentences, paragraphs, articles, etc. Co-occurrences between person categories and evaluations within counting units of sentences or smaller text units indicate explicit discriminations. Co-occurrences which occur within of a larger text unit indicate implicit or latent discriminations (see more in [Galliker *et al.*, 1997]. The distribution of co-occurrences and their movement in time can be shown by time-space projections of the items (further details in [Galliker, 1996]. The relevant frequencies (e.g. for a study on migrants and asylum-seekers) can be brought into relation with one another and compared with the frequency of person categories that serve as control groups with a statistical comparison (e.g. ethnic compatriots from a friendly neighboring state).

It is possible to empirically investigate the relation of open and subtle devaluations of 'foreigners' with other groups at the time of certain political events. For example, the explicit devaluations of migrants decreased at the time that a confederate anti-racism commission was initiated, while the latent ones, at least temporarily, increased [Galliker *et al.*, 1998]. How decisive the contiguity operationalized by the co-occurrence is for the development of the discourse could be determined in most of the investigations, whereas the issue of whether or not devaluations are negated is in any case of secondary significance (compare [Galliker *et al.*, 1997]).

This study also shows that co-occurrence analysis is not a purely quantitative method, but allows for transitions to more qualitative analyses. This procedure makes it possible to do more than just determine the fundamental structure of discrimination (categorization and evaluation). Differentiations can be made at any time, for example by counting negative and positive evaluations of 'outgroups' and 'ingroups' (young people) with regard to various counting units (e.g. [Galliker and Klein, 1997]). Special text units which contain three or more verbal occurrences could be isolated, making classifications with regard to content possible in context of actual political events (compare [Galliker *et al.*, 1997; Galliker *et al.*, 1998]). One can also use co-occurrence analysis to verify the assumption of this study, namely that sentences with person categories which have a devaluative connotation only on the basis of their context have a more than coincidentally high frequency of particles and modal words, indicating their implicitly discriminating character [Wagner, 1997; Wagner *et al.*, 1997].

In contrast to the traditional method of counting isolated words based exclusively on the text corpus as a whole, with the following cluster analysis in regard to this particular corpus, co-occurrence analysis preserves the psychologically relevant details by counting word combinations per text units of various sizes. The verbal distribution and their movement in time can also be shown. Thus co-occurrence analysis proves useful as a method to see patterns and trends in verbal contexts. The use of CD-ROMs as a data source for co-occurrence analysis will improve accessibiblity and thereby make it possible for virtually anyone to check and test the results. Since the operations necessary for the analysis are conducted using a computer program on the entire text in an identical manner and can be repeated at any time by different individuals, there are are no problems of objectivity and reliability. Concerning the scope of a co-occurrence analysis one must conclude that the results are only valid for the verbal context (e.g. three years of back-issues of a particular newspaper). This scope cannot be expanded by the mere statistical procedure of generalizing, but it can be successively broadened or bundled according to greater or lesser change in the starting conditions (newspapers of various circulation sizes, political orientations, and country sizes etc.) by further research on implicitness in other contexts (for more on methodology see [Foppa, 1986]).

Psychologisches Institut der Universität Heidelberg, Germany.

REFERENCES

[Berger and Bradac, 1982] Ch. R. Berger and J. J. Bradac. *Language and Social Knowledge: Uncertainty in Interpersonal Relations.* Edward Arnold, London, 1982.

[Billig, 1988] M. Billig. Social representation, anchoring and objectification: A rhetorical analysis. *Social Behavior*, **3**, 1–16, 1988.

[Billig, 1993] M. Billig. Studying the thinking society: Social representations, rhetoric, and attitudes. In *Empirical Approaches to Social Representations* G. M. Breakwell and D. V. Canter, eds. pp. 39–62. Clarendon Press, Oxford, 1993.

[Bronfenbrenner, 1979] U. Bronfenbrenner. *The Ecology of Human Development: Experiments by Nature and Design.* Harvard University Press, Cambridge, MA, 1979.

[de Certeau, 1988] M. de Certeau. *The Practice of Everyday Life*, University of Carlifornia Press, Berkeley, 1988.

[Dovidio and Garetner, 1985] J. F. Dovidio and S. L. Gaertner. *Prejudice, Discrimination, and Racism.* Orlando, FL, 1985.

[Fowler, 1983] H. W. Fowler. *A Dictionary of Modern English Usage.* Oxford University Press, Oxford, 1983..

[Foppa, 1986] K. Foppa. 'Typische Fälle' und der Geltungsbereich empirischer Befunde. *Swiss Journal of Psychology*, **45**, 151–163, 1986.

[Forgas, 1988] J. P. Forgas. Episode representations in intercultural communication. In *Theories in Intercultural Communication*, Y. Y. Kim and W. B. Gudykunst, eds. pp. 186–212, Sage, Newbury Park, CA, 1988.

[Frege, 1892/1994] G. Frege. Ueber Sinn und Bedeutung. In *Gottlob Frege: Funktion, Begriff, Bedeutung*, G. Patzig, ed., pp. 40–65. Vandenhoeck & Ruprecht, Göttingen, 1892–1994.

[Galliker, 1977] M. Galliker. *Müssen wir uns auf das Sprechen vorbereiten? Ein genetisch-praktischer Ansatz der Psycholinguistik auf der Grundlage von Wittgenstein.* Haupt, Bern, 1977.

[Galliker, 1980] M. Galliker. *Arbeit und Bewusstsein. Eine dialektische Analyse von Gesprächen mit Arbeitern, Angestellten, Beamten und Selbständig Erwerbenden.* Campus, Göttingen, 1980.

[Galliker, 1990] M. Galliker. *Sprechen und Erinnern.* Hogrefe, Göttingen, 1990.

[Galliker, 1996] M. Galliker. Delegitimierung von Migranten im Mediendiskurs. *Kölner Zeitschrift für Soziologie und Sozialpsychologie*, **48**, 704–727, 1996.

[Galliker *et al.*, 1996] M. Galliker, J. Herman, F. Wagner and D. Weimer. Co-Occurrence-Analysis von Medientexten. *Medienpsychologie*, **8**, 3–20, 1996.

[Galliker *et al.*, 1998] M. Galliker, J. Herman, K. Imminger and D. Weimer. The investigation of contiguity: Co-occurrence analysis of print media using CD-ROMs as a new data source, illustrated by a discussion on migrant delinquency in a daily newspaper. *Journal of Language and Social Psychology*, **17**, 200–217', 1998.

[Galliker *et al.*, 1994] M. Galliker, M. Huerkamp, F. Wagner and C. E. Graumann. Validierung eines facettentheoretischen Modells sprachlicher Diskriminierung anhand von Beurteilungen deutscher und ausländischer Probanden. *Sprache & Kognition*, **13**, 203–220, 1994.

[Galliker *et al.*, 1995] M. Galliker, M. Huerkamp and F. Wagner. The social perception and judgement of foreigners. In *Perception—Evaluation—Interpretation*, B. Boothe, R. Hirsig, A. Helminger, B. Meier and R. Volkart, eds. pp. 134–140. Hogrefe, Seattle, 1995.

[Galliker *et al.*, 1997] M. Galliker, K. Imminger, D. Weimer and H. Bock. Intensivierung des Diskurses durch Verneinung: Quantitative und qualitative Analyse der Co-Occurrences von 'Soldaten' und 'Mörder' in der FAZ im Vergleich mit der NZZ. In *Evidenzen im Fluss*, A. Disselnkötter, S. Jäger, H. Kellershohn and S. Slobodzian, eds. pp. 99–120. DISS, Duisburg, 1997.

[Galliker and Klein, 1997] M. Galliker and M. Klein. Implizite positive und negative Bewertungen: Eine Kontextanalyse der Personenkategorien 'Senioren', 'ältere Menschen', 'alte Menschen' und 'Greise' bei drei Jahrgängen einer Tageszeitung. *Zeitschrift für Gerontopsychologie und -psychiatrie*, **10**, 27–41, 1997.

[Galliker and Wagner, 1995] M. Galliker and F. Wagner. Implizite Diskriminierungen und Antidiskriminierungen anderer Menschen im öffentlichen Diskurs. *Zeitschrift für Politische Psychologie*, **3**, 69–86, 1995.

[Galliker *et al.*, 1995] M. Galliker, D. Weimer and F. Wagner. The contribution of Facet Theory to the interpretation of findings: Discussing the validation of the basic facets of the model of verbal discrimination. In *Facet Theory: Analysis and Design*, J. J. Hox, G. J. Mellenbergh, and P. G. Swanborn, eds., pp. 107-117. Setos, Zeist, 1995.

[Giles and Coupland, 1991] H. Giles and N. Coupland. *Language: Contexts and consequences*. Open University Press, Buckingham, 1991.

[Graumann, 1995] C. F. Graumann. Discriminatory discourse. *Patterns of Prejudice*, **29**, 69–83, 1995.

[Graumann and Wintermantel, 1989] C. F. Graumann and M. Wintermantel. Discriminatory speech acts: A functional approach. In *Stereotyping and Prejudice: Changing Conceptions*, D. Bar-Tal, C. F. Graumann, A. W. Kruglanski and W. Stroebe, eds. pp. 184-204. Springer-Verlag, New York, 1989.

[Gumperz, 1982] J. J. Gumperz. *Language and Social Identity*. Cambridge University Press, 1982.

[Gumperz, 1992] J. J. Gumperz. Contextualization and understanding. In *Rethinking context*, A. Duranti and C. Goodwin, eds. pp. 229–252. Cambridge University Press, New York, 1992.

[Helbig, 1990] G. Helbig. *Lexikon deutscher Partikeln*. VEB Verlag Enzyklopädie, Leipzig, 1990.

[Helbig and Helbig, 1993] G. Helbig and Helbig. *Lexikon deutscher Modalwörter*. Langenscheidt, Liepzig, 1993.

[Hughes, 1995] R. Hughes. *Cultures of complaint*. Oxford University Press. 1993.

[Jäger, 1993] S. Jäger. *Kritische Diskursanalyse*. DISS, Duisburg, 1993.

[Lang, 1992] A. Lang. Kultur als 'externe Seele': eine semiotisch-ökologische Perspektive. In *Psychologische Aspekte des kulturellen Wandels*, Chr. G. Allesch, E. Billmann-Mahecha and A. Lang, eds. pp. 11–32. Wien: Verlag des Verbandes der wissenschaftlichen Gesellschaften Österreichs, 1992.

[Luutz, 1994] W. Luutz. *'Das soziale Band ist zerrissen': Sprachpraktiken sozialer Desintegration*. Universitätsverlag, Leipzig, 1994.

[Markov'a, 1990] I. Markovà. A three-step process as a unit of analyis in dialogue. In *The Dynamics of Dialogue*, I. Markovà and K. Foppa, eds. pp. 129–146. Harvester Wheatsheaf, Hemel Hempstead, 1990.

[Pettigrew, 1989] T. E. Pettigrew. The nature of modern racism in the United States. *Revue Internationale de Psychologie Sociale*, **2**, 291–303, 1989.

[Pettigrew and Meertens, 1995] T. E. Pettigrew and R. Meertens. Subtle and blatant prejudice in western Europe. *European Journal of Social Psychology*, **25**, 57–75, 1995.

[Sinclair, 1992] J. Sinclair. *Collins Cobuild. English usage*. Harper, London, 1992.

[Strauss, 1991] A. L. Strauss. *Grundlagen qualitativer Sozialforschung*. Fink, München, 1991.

[van Dijk, 1985] T. A. van Dijk, ed. *Handbook of Discourse Analysis (Vol. 4): Discourse Analysis in Society*. Academic Press, London, 1985.

[van Dijk, 1991] T. A. van Dijk. *Racism and the Press*. Routledge, London, 1991.

[van Dijk, 1994] T. A. van Dijk. Principles of critical discourse analysis. *Discourse & Society*, **4**, 249–283, 1994.

[Walvin, 1992] J. Walvin. *Black Ivory: A History of British Slavery*. Harper Collins, Glasgow, 1992.

[Wagner, 1997] F. Wagner. *Implizite Sprache: Diskriminierung als Sprechakt. Lexikalische Indikatoren sprachlicher Implizitheit*. Dissertation. University of Heidelberg, 1997.

[Wagner *et al.*, 1997] F. Wagner, M. Huerkamp, M. Galliker and C. F. Graumann. Implizite sprachliche Diskriminierung aus linguistischer Sicht. In *Dialoganalyse V*, E. Pietri, ed. pp. 529–536. Niemeyer, Tübingen, 1997.

[Wagner *et al.*, 1993] F. Wagner, H. Huerkamp, H. Jockisch and C. F. Graumann. Sprachliche Diskriminierung. In *Dialoganalyse IV: Referate der 4. Arbeitstagung Basel 1992*, H. Löffler, ed. pp. 281–288. Niemeyer, Tübingen, 1993.

[Wittgenstein, 1971] L. Wittgenstein. *Vorlesungen und Gespräche über Ästhetik, Psychologie und Religion*. Vandenhoeck & Ruprecht, Göttingen, 1971.

[Wittgenstein, 1980] L. Wittgenstein. *Remarks on the Philosophy of Psychology* (Vol. 2). Blackwell, Oxford, 1980.

[Züll *et al.*, 1991] C. Züll, P. Ph. Mohler and A. Geis. *Computerunterstützte Inhaltsanalyse mit TEXTPACK PC*. Gustav Fischer, Stuttgart, 1991.

ALESSANDRO CIMATTI AND LUCIANO SERAFINI

A CONTEXT-BASED MECHANIZATION OF MULTI-AGENT REASONING

1 INTRODUCTION AND MOTIVATIONS

Belief contexts [Giunchiglia, 1993; Giunchiglia and Serafini, 1994; Giunchiglia *et al.*, 1993] are a formalism for the representation of propositional attitudes. Their basic feature is *modularity*: knowledge can be distributed in different, separated modules, called contexts; the interactions between them, i.e. the transfer of knowledge between contexts, can be formally defined according to the application. For instance, the beliefs of an agent can be represented with one or more contexts, distinct from the contexts representing beliefs of other agents; different contexts can be used to represent the beliefs of an agent in different situations. Interaction between contexts can express the effect of communication between agents, and the evolution of their beliefs (e.g. learning, belief revision).

In [Cimatti and Serafini, 1995] and [Cimatti and Serafini, 1996], we discuss how belief contexts can be used to formalize multi-agent reasoning. In this paper we address the issue of the *mechanization* of multi-agent reasoning. We show that using belief context also gives implementational advantages.

First, the representation formalism is *incremental*: adding new contexts to the system has no effects other than the ones which are explicitly encoded in the interaction mechanisms. This allows for incremental and independent development of different parts of the knowledge base; furthermore, it opens up the possibility to parallelize the inference process in a natural way.

Second, substantial parts of the reasoning process are *local* to modules. For instance, in order to represent the "internal" reasoning of an agent it is not necessary to take into account the information describing a different agent. Local reasoning is captured by general purpose reasoning inside a well defined, simple module rather than by ad hoc reasoning techniques, which try to isolate the relevant information in a global, unstructured theory.

Finally, non-local reasoning steps, corresponding to interaction between modules, often have a very *natural* interpretation. The advantage is that the search strategy can exploit the structure of the system, which can be tailored to the structure of the problem: therefore, the process of inference is much more efficient, being problem dependent, than a uniform and problem independent strategy.

The goal of this paper is to show how the formal features of belief contexts give advantages on the mechanization. We do this by discussing the

P. Bonzon, M. Cavalcanti and R. Nossum (eds.), Formal Aspects of Context, 65–83.

mechanization in GETFOL of a paradigmatic scenario of multi-agent reasoning (GETFOL [Giunchiglia, 1994] is an interactive system for the mechanization of multicontext systems, developed on top of a reimplementation [Giunchiglia and Weyhrauch, 1991] of FOL [Weyhrauch, 1980]). Our scenario is the following [McCarthy, 1990]:

> "A certain King wishes to test his three wise men. He arranges them in a circle so that they can see and hear each other and tells them that he will put a white or black spot on each of their forehead but that at least one spot will be white. In fact all three spots are white. He then repeatedly asks them: "Do you know the color of your spot?". What do they answer?"

Although this puzzle might be thought of as a toy example, the reading presented here forces us to formalize issues such as multiagent belief, common and nested belief, ignorance and ignorance ascription. These are the problems that a real system for reasoning about propositional attitudes (e.g. [Creary, 1979; Haas, 1986]) has to face.

The paper is structured as follows. In section 2 we give an overview of belief contexts. In section 3 we show how belief contexts are mechanized in GETFOL. In section 4 we discuss the mechanized solution to the puzzle. In section 5 we draw some conclusions. A comparison with related works can be found in [Cimatti and Serafini, 1995].

2 BELIEF CONTEXTS

Intuitively, a (belief) context represents a collection of beliefs under a certain point of view. Different contexts may be used to represent the belief sets of different agents about the world. In the three wise men scenario (TWM) there are three agents (wise men 1, 2 and 3), with certain beliefs about the state of the world. The context of the first wise man contains the fact that the spots of the second and third wise men are white, that the second wise man believes that the spot of the third wise man is white, and possibly other information. Other contexts may formalize a different view of the world, e.g. the set of beliefs that an agent ascribes to another agent. For example, the (context formalizing the) set of beliefs that 1 ascribes to 2 contains the fact that the spot of 3 is white; however, it does not contain the fact that his own (i.e. 2's) spot is white, because 2 can not see his own spot. A context can also formalize the view of an observer external to the scenario (e.g. us, or even a computer, reasoning about the puzzle). This context contains the fact that all the spots are white, and also that each of the agents knows the color of the other spots, but not that he knows the color of his own.

Formally, a context is a theory which we present as a formal system $\langle L, \Omega, \Delta \rangle$, where L is a logical language, $\Omega \subseteq L$ is the set of axioms (basic sensor systems are collected. Since a user can be at only one place at a time, coordinates gathered at the same time must be from the same physical location.

The collected coordinates are specific cases of how *(x, y)* maps to *(x', y')* in Equation (1). With three such mappings, we can compute the transformation matrix, since the equation is actually two linear equations with three variables each. Using only three such mappings will probably result in large errors because of errors in the detected coordinates, so instead we use two-dimensional multiple linear regression with a large number of mappings. Our framework gathers mappings from coordinates in the newly deployed location sensor system to coordinates in the location manager until the rate of change in the computed transformation matrix falls below a specific threshold.

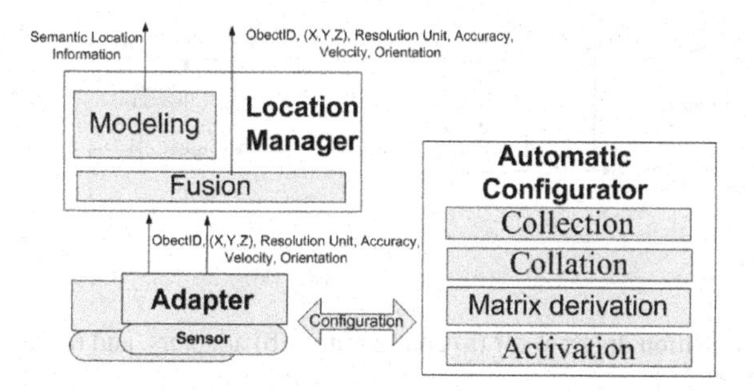

Fig. 1. Overall architecture of the proposed framework

Once the transformation matrix is obtained, the framework sends the result to the adapter, which will use the transformation matrix to transform sensor-specific coordinates to reference coordinates when reporting locations to the location manager. Fig. 1 illustrates the architecture of our framework.

3.2 System Components

As described in Section 3.1, the framework consists of three major components – location manager, configurator, and adapters.

The location manager gathers data from various location sensor systems and tracks physical objects continuously. It performs sensor fusion in order to obtain more accurate locations. It also reports locations either symbolically or as physical coordinates according to application needs. The location manager is basically the central clearinghouse from which applications obtain locations of physical objects.

The configurator supports the automatic configuration of adapters so that they can integrate with the location manager. Some of the parameters that are configured are the transformation matrix and adapter identifier. In order to acquire the transformation matrix, the configurator has collection, collation, matrix derivation, and activation components. The collection component collects coordinates from the newly deployed location sensor system and the location manager. The collation component binds coordinates collected at the same time. The matrix derivation component computes the transformation matrix based on these coordinates. The activation component is responsible for configuring the transformation matrix in the adapter.

The adapter enables plug-and-play deployment of heterogeneous location sensors. Thanks to the adapter, the upper layers such as the location manager need not consider the diversity in underlying sensor technologies. It includes a transformation matrix in order to integrate with the location manager. The adapter includes error attributes which influence how the transformation matrix is obtained. An adapter also includes the estimated error when reporting locations.

3.3 Configuration and Integration Protocol

In this section we describe the protocol between the system components for configuring and integrating a new location sensor system.

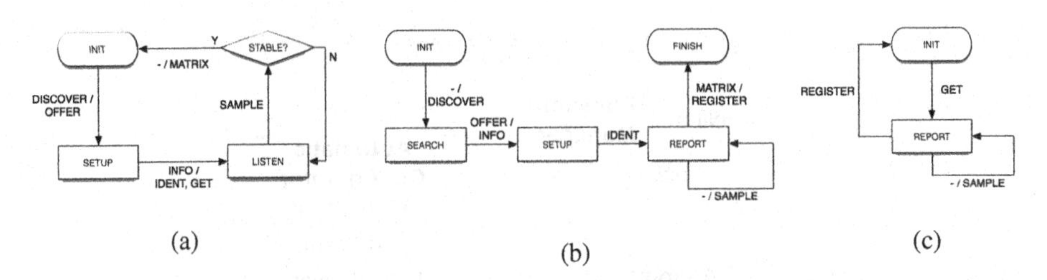

Fig. 2. State-transition diagram for (a) configurator, (b) adapters, and (c) location manager

Figure 2(a) illustrates the state-transition diagram for the configurator. The configurator starts out in the INIT state, and waits for a DISCOVER message from an adapter. Once it receives a DISCOVER message, it replies with an OFFER message, which includes the network location of the configurator, and changes to the SETUP state. In the SETUP state, the configurator waits for an INFO message from the adapter, which includes information about the sensor system such as its type, reliability, precision and coverage. It then allocates an identifier for the adapter and sends it within an IDENT message to the adapter. Then IDENT message also includes the network location of the location manager. It also sends a GET message to the location manager. The configurator then changes to the LISTEN state, during which it continuously receives SAMPLE messages from both the adapter and the location manager. The SAMPLE messages include the coordinates of the designated sensor tags and are the adapter identifier they come from. These are gathered until enough data is collected so that the computed transformation matrix or location is stable for ranged sensors or point sensors, respectively. Once the configurator has determined that enough data has been collected, it sends a MATRIX message to the adapter, which includes the transformation matrix that should be used by the adapter, and returns to the INIT state where it waits for other new location sensor systems.

Figure 2(b) illustrates the state-transition diagram for adapters. Starting out in the INIT state, an adapter broadcasts a DISCOVER message over the network and changes to the SEARCH state. The configurator will respond with an OFFER message, which the adapter replies to with an INFO message and then changes to the SETUP state. The adapter waits for an IDENT message from the configurator in the SETUP state, from which it will learn its identifier and the network location of the location manager. It then switches to the REPORT state, where it continuously sends the coordinates of the designated sensor tag from the new location sensor system to the configurator in SAMPLE messages. It continues this until it receives a MATRIX message from the configurator. The adapter uses this message to configure its transformation matrix. It then registers its identifier, network location and information about the sensor system with the location manager using a REGISTER message, at which point configuration of the adapter is complete and it can start reporting locations to the location manager using the reference coordinate system.

Figure 2(c) illustrates the state-transition diagram for the location manager. The location manager starts out in the INIT state, where it waits for a GET message from the configurator. Once this message is received, it starts reporting the coordinates of the designated sensor tag from an existing location sensor system using SAMPLE messages. It continues this until it receives a REGISTER message from the adapter. (SAMPLE messages sent from the location manager to the configurator after the configurator sends a MATRIX message but before the adapter sends a REGISTER message are discarded by the configurator.) Integration of the new location sensor system is complete at this point, and the location manager can start measuring locations and apply sensor fusion using the information in the REGISTER message.

3.4 User Actions

Our framework currently supports two methods for deploying a new location sensor system. In both methods, the user carries sensor tags designated by the configurator from both the location sensor system being deployed and a location sensor system already integrated with the location manager.

In the first method, which we will call the guideless configuration method, the user tells the adapter that configuration and integration should begin using a PDA. The user then randomly moves around the area covered by the location sensor system being deployed. Location coordinates are continuously and automatically gathered using the carried sensor tags, which the configurator uses to compute the transformation matrix. Once the configurator determines that it has gathered enough data, it continues with the configuration and integration procedure after notifying the user that he can stop moving via the PDA.

The second method, which we will call the guided configuration method, starts out similarly with the user starting the configuration with the PDA. However, an application on the PDA guides the actions of the user. The application uses a map viewer to suggest to the user a specific location to move to. The user then moves to the approximate location, waits several seconds, and via the PDA tells the adapters to report a fixed number of coordinates to the configurator. This repeats until the configurator determines that enough data has been gathered.

The first method is much more convenient for users in that no interaction with the configurator is required while coordinates are gathered. However, the second method can be more accurate because some sensor systems become more inaccurate when there is movement (e.g. the error in Ubisense is several times larger when walking compared to standing still). Using the map viewer to suggest locations to move to also helps ensure that coordinates are gathered from a widely dispersed range, which can help improve the accuracy of the derived transformation matrix, although this requires that a map of the area be available.

4 Evaluation

4.1 Test Environment

We implement our framework as part of the Active Surroundings environment, our ubiquitous computing middleware [5]. We test the proposed location framework in a testbed which is comprised of a $36m^2$ room and includes a variety of sensors, hardware appliances, and software systems necessary for a ubiquitous computing environment.

Among the sensors included is the Ubisense location system [10]. Ubisense detects locations using fixed sensors installed inside the room. These sensors receive signals from a user-carried radio beacon and estimate the location of the beacon from the relative signal strengths. Our installation of Ubisense, which only covers the Active Surroundings environment, exhibits an error range of 25cm for non-moving beacons, although the error range grows several times larger while beacons are moving.

We also develop a WLAN-based location system for the Active Surroundings environment. It uses the signal strengths of multiple wireless access points to estimate location [8]. It covers the entire building but exhibits an error range of 3m.

We test a scenario where a Ubisense location sensor system is being installed into an environment with an already integrated WLAN-based location sensor system. We test manual configuration, guideless configuration and guideless configuration for deploying Ubisense. Each test is repeated five times.

The reference coordinate system is set to that used by the WLAN-based location sensor system. The origin of this coordinate system is set to one of the corners of the room, and the axes are oriented parallel to the walls of the room. The reference coordinate system is two-dimensional, so height is not represented.

4.2 Results and Analysis

We measure the amount of time to complete configuration for each test. We also measure how much error that is exhibited in each physical location when measuring locations with the Ubisense location sensor system after it is configured and integrated. Sensor fusion is not used when measuring the errors.

The average amount of time taken to complete manual configuration is about 20 minutes. The amount of skill required for manual configuration suggests that it would be difficult and more time consuming for casual users to manually configure and integrate a new location sensor system.

The average amount of time taken to complete guideless configuration is about 5 minutes. The configurator collects 375 mappings of sensor-specific coordinates to reference coordinates during this time. The average amount of time taken to complete guided configuration is about 10 minutes. The configurator collected 15 samples each in 25 locations during this time. The resulting error distributions of the configured Ubisense system are shown in Figures 3, 4, and 5 for manual, guideless, and guided configuration.

facts of the view), and Δ is a deductive machinery. This general structure allows for the formalization of agents with different expressive and inferential capabilities [Giunchiglia *et al.*, 1993]. We consider belief contexts where Δ is the set of classical natural deduction inference rules [Prawitz, 1965], and L is described in the following. To express statements about the spots, L contains the propositional constants W_1, W_2 and W_3. W_i means that the spot of i is white. To express belief, L contains well formed formulas (wff) of the form $B_i(\text{"}A\text{"})$, for each wff A and for $i = 1, 2, 3$. Intuitively, $B_i(\text{"}A\text{"})$ means that i believes the proposition expressed by A; therefore, $B_2(\text{"}W_1\text{"})$ means that 2 believes that 1 has a white spot. The formula $CB(\text{"}A\text{"})$, with A being a formula, means that the proposition expressed by A is a *common belief*, i.e. not only all the agents believe it, but also they believe it to be a common belief (see for instance [Moore, 1982]). For instance, we express that at least one of the spots is white is a common belief with the formula $CB(\text{"}W_1 \vee W_2 \vee W_3\text{"})$.

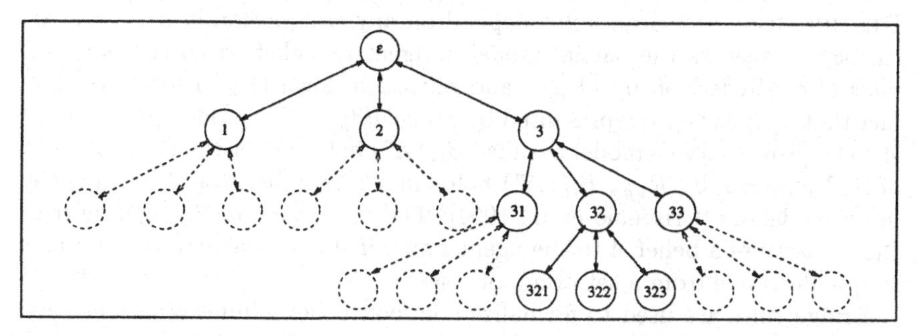

Figure 1. The context structure to express multiagent nested belief

Contexts are organized in a tree (see figure 1). We call ϵ the root context, representing the external observer point of view; we let the context i formalize the beliefs of wise man i, and ij the beliefs ascribed by i to wise man j. Iterating the nesting, the belief context ijk formalizes the view of agent i about j's beliefs about k's beliefs. In general, a finite sequence of agent indexes, including the null sequence ϵ, is a context label, denoted in the following with α.

The interpretation of a formula depends on the context we consider. For instance, the formula W_1 in the external observer context, written $\epsilon : W_1$ to stress the context dependence, expresses the fact that the first wise man has a white spot. The same formula in context 232, i.e. $232 : W_1$, expresses the (more complex) fact that 2 believes that 3 believes that 2 believes that 1 has a white spot. Notice that "2 believes that 3 believes that 2 believes that.." does not need to be stated in the formula. Indeed, context 232 represents the beliefs that 2 believes to be ascribed to himself by 3. However, it would need to be made explicit if the same proposition were expressed in the context of the external observer ϵ: the result is the (more complex) formula $B_2(\text{"}B_3(\text{"}B_2(\text{"}W_1\text{"})\text{"})\text{"})$. This shows that a fact can be expressed with belief contexts in different ways.

Contexts are "glued together" in multi-context systems by means of *bridge rules* [Giunchiglia, 1993], i.e. rules with premises and conclusions in distinct belief contexts. Bridge rules are a general tool for the formalization of interactions between contexts: the derivability of a formula in a context can yield the derivability of another formula in another context.

A very important case of interaction between contexts is the following: $232 : W_1$ has to be provable if and only if $\epsilon : B_2(\text{"}B_3(\text{"}B_2(\text{"}W_1\text{"})\text{"})\text{"})$ is, as they have the same meaning. This is formalized by the following bridge rules, called *reflection* rules [Giunchiglia and Serafini, 1994]:

$$\frac{\alpha : B_i(\text{"}A\text{"})}{\alpha i : A}\mathcal{R}_{dn.} \qquad \frac{\alpha i : A}{\alpha : B_i(\text{"}A\text{"})}\mathcal{R}_{up.}$$

RESTRICTION: $\alpha i : A$ does not depend on any assumption in αi. Context αi may be seen as the partial model of agent i's beliefs from the point of view of α. Reflection up ($\mathcal{R}_{up.}$) and reflection down ($\mathcal{R}_{dn.}$) formalize the fact that i's beliefs are represented by ***provability*** in this model. $\mathcal{R}_{up.}$ forces A to be provable in i's model because $B_i(\text{"}A\text{"})$ holds under the point of view of α. Viceversa, by $\mathcal{R}_{up.}$, $B_i(\text{"}A\text{"})$ holds in α's view because A is provable in his model of i [Criscuolo *et al.*, 1994]. The restriction on $\mathcal{R}_{up.}$ guarantees that α ascribes a belief A to the agent i only if A is provable in αi, and not simply derivable from a set of hypotheses.

Bridge rules are used to formalize common belief [Giunchiglia and Serafini, 1991]. The bridge rule CB_{inst} allows us to derive belief of a single agent from common belief, i.e. to ***inst***antiate common belief. The bridge rule CB_{prop} allows us to derive, from the fact that something is a common belief, that an agent believes that it is a common belief, i.e. to ***prop***agate common belief.

$$\frac{\alpha : CB(\text{"}A\text{"})}{\alpha i : A}CB_{inst} \qquad \frac{\alpha : CB(\text{"}A\text{"})}{\alpha i : CB(\text{"}A\text{"})}CB_{prop}$$

Finally, a bridge rule is used to reason about ignorance, i.e. to infer formulas of the form $\neg B_i(\text{"}W_i\text{"})$. We know that belief corresponds to provability in the context modeling the agent. However, since this model is partial, non belief does not correspond to simple non provability. Intuitively, we relate ignorance to non-derivability, rather than non-provability, as follows: infer that agent i does not know A, if A can not be derived in the context modeling i from those beliefs of i explicitly stated to be relevant. Formally, all we need are ***relevance statements*** and a bridge rule of ***belief closure***. A relevance statement is a formula of L of the form $ARF_i(\text{"}A_1,\ldots,A_n\text{"}, \text{"}A\text{"})$, where $A_1,\ldots,A_n,A$ are formulas of L. The meaning of $ARF_i(\text{"}A_1,\ldots,A_n\text{"}, \text{"}A\text{"})$

is that $A_1, \ldots, A_n$ are all the relevant facts available to i to infer the conclusion A. The bridge rule of belief closure, which allows us to infer ignorance, is the following:

$$\frac{\alpha i : A_1 \cdots \alpha i : A_n \quad \alpha : ARF_i(\text{"}A_1, \ldots, A_n\text{"}, \text{"}A\text{"})}{\alpha : \neg B_i(\text{"}A\text{"})} \text{ Bel-Clo}$$

RESTRICTIONS: $A_1, \ldots, A_n \not\vdash_{\alpha i} A$; $\alpha i : A_1, \ldots, \alpha i : A_n$ do not depend on any assumption in αi. A detailed discussion of this rule is out of the scope of this paper (but see [Cimatti and Serafini, 1995; Cimatti and Serafini, 1996]). Here we only point out that the main advantage of our solution with respect to other mechanisms, e.g. circumscriptive ignorance [Konolige, 1986], is expressivity. We can express relevance hypotheses on the knowledge of an agent *in* the formal language, rather than leaving them unspoken at the informal metalevel.

`Whitei`	$\xrightarrow{sdn}$	`"Whitei"`
		Where `Whitei` is `White1`, `White2` or `White3`
`P(t)`	$\xrightarrow{sdn}$	`apply("P", encode(t))`
		Where `P` is `B1`, `B2` or `B3` and `t` is a term
`P(s,t)`	$\xrightarrow{sdn}$	`apply2(`*sdn*`(P), encode(s), encode(t))`
		Where `P` is `ARF1`, `ARF2` or `ARF3` and `s` and `t` are terms
`not A`	$\xrightarrow{sdn}$	`makenot(`*snd*`(A))`
		Where `A` is a wff
`A and B`	$\xrightarrow{sdn}$	`makeand(`*snd*`(A), `*snd*`(B))`
		Where `A` and `B` are wffs
`A or B`	$\xrightarrow{sdn}$	`makeor(`*snd*`(A), `*snd*`(B))`
		Where `A` and `B` are wffs
`A imp B`	$\xrightarrow{sdn}$	`makeimp(`*snd*`(A), `*snd*`(B))`
		Where `A` and `B` are wffs

Figure 2. The structural descriptive naming function

3 MECHANIZING BELIEF CONTEXTS

In this section we describe how the MC system described in the previous section is mechanized in GETFOL. In next section we show how (different formulations of) the puzzle are solved in this mechanization. For the sake of simplicity we suppose that the wise men don't answer simultaneously, and that wise 1, 2 and 3 speak in numerical order. In [Cimatti and Serafini, 1995] we formalize the reasoning of the wise men in three situations (i.e.

before the first, the second and the third answer) with three different systems of contexts, all having the same structure shown in figure 1. For the mechanization we collect the three trees of contexts in a unique MC system (called TWM). The context Si mechanizes the external observer context ϵ of the MC system for the i-th situation. Analogously, the TWM context Si_j mechanizes the context j for the i-th situation, and so on.

We start by declaring the contexts of TWM. The GETFOL command to define a context is **makecontext**. Contexts can be grouped to form an MC system by using the command **makemcontext**. By the statements (GETFOL:: is the GETFOL prompt):

```
GETFOL:: makecontext S1 S1_1 S2 S2_2 S2_3
                     S3 S3_3 S3_32 S3_321;
GETFOL:: makemcontext TWM S1 S1_1 S2 S2_2 S2_3
                          S3 S3_3 S3_32 S3_321;
```

TWM is declared as an MC system composed of the contexts S1, S1_1, S2, S2_2, and so on. The modularity and monotonicity of belief contexts [Cimatti and Serafini, 1995] enable us to implement contexts lazily, i.e. contexts are created and added to the MC system on line, only when it is required to use them. This implies that TWM can be suitably extended, if necessary. The advantages of laziness are twofold: first, it allows for a more efficient implementation; second, it is always possible to mechanize the relevant part an infinite structure of contexts with a finite data structure.

We declare now the language L defined in previous section. The command **declare** (options **sentconst**, **predconst**, **indconst**, **funconst**) allows us to declare sentential, predicate, individual and functional symbols, respectively.

```
GETFOL:: declare sentconst White1 White2 White3;
GETFOL:: declare predconst B1 B2 B3 CB 1;
GETFOL:: declare predconst ARF1 ARF2 ARF3 2;
```

The language L contains the infinite set of constants, "A" where A is a wff, which are the *quotation mark names* of the wffs. As we cannot declare an infinite set of symbols, we have to provide a finite presentation for such a set. For this reason we use *structural descriptive names* [Tarski, 1956; Giunchiglia and Traverso, 1996] in place of quotation mark names. We declare constants to name the basic components of wffs (i.e. the symbols declared above), and functions to construct the name for a complex formula from the names of its subexpressions.

```
GETFOL:: declare indconst ~White1~ ~White2~
                          ~White3~;
GETFOL:: declare indconst ~B1~ ~B2~ ~B3~ ~CB~;
GETFOL:: declare indconst ~ARF1~ ~ARF2~ ~ARF3~;
```

Intuitively ~White1~ is the name of the formula White1.

```
GETFOL:: declare funconst makenot 1;
GETFOL:: declare funconst makeand makeor
                          makeimp 2;
GETFOL:: declare funconst apply 2;
GETFOL:: declare funconst apply2 3;
GETFOL:: declare funconst encode 1;
```

makenot, makeand, makeor, makeimp, apply and apply2 are the functions used to define structural descriptive names. For instance, the name of the wff White1 and not White2 is makeand(~White1~,makenot(~White2~)). The function symbol encode is necessary to build names for wffs which contain names of other wffs. Intuitively encode formalizes the function that maps a wff into its structural descriptive name. This function, called *sdn*, is defined in figure 2. Intuitively, if t_w is a structural descriptive name for the wff w, then encode(t_w) is the structural descriptive name for t_w. For instance, encode(~White1~) is the name of the constant ~White1~. The name of an atomic wff of the form B1(t_w) is obtained by applying apply to the name of the predicate symbol B1 (i.e. ~B1~) and to the *name* of t_w (i.e. encode(t_w)). For instance, the name of B1(a) is apply(~B1~,encode(a)).

The language defined above can be copied to all the contexts of TWM by the GETFOL command copylex. As all contexts have the same language, we can interpret ~A~ as the structural descriptive name of the wff A in all the contexts. This assumption is satisfactory for the work presented here. However, considering the case of contexts with different languages requires to relax this assumption, and introduces further philosophical and technical issues which are out of the scope of the paper.

In GETFOL every context has (by default) a set of first-order natural deduction style inference rules. Bridge rules can be configured according to the application. We impose the tree structure of figure 1 on the contexts by means of the GETFOL command newcommand, which allows us to declare new inference rules by instantiating a number of given rule schemata. For instance, to declare reflection down from S1 to S1_1, we have to switch into conclusion's context S1_1 and instantiate the reflection down schema RDOWNPROP by specifying both the premise's context (i.e. S1) and the predicate on which reflection down works (i.e. B1). The other reflection rules can be declared analogously. In the following we give some examples:

```
GETFOL:: switchcontext S1_1;
GETFOL:: newcommand rdown IS RDOWNPROP S1 B1;
GETFOL:: newcommand cb_inst IS RDOWNPROP S1 CB;
GETFOL:: newcommand cb_prop IS RDOWNCB2 S1 CB;
GETFOL:: newcommand rdownall IS RDOWNALL S1 B1 CB;

GETFOL:: switchcontext S1;
```

```
GETFOL:: newcommand belclo_a1 IS BELCLO S1_1 B1 ARF1;

GETFOL:: switchcontext S2;
GETFOL:: newcommand rup IS RUPPROP  S2_2 B2;
```

rdownall mechanizes a derived bridge rule, which considers the atomic formulas of the context S1 which can affect (either via $\mathcal{R}_{dn.}$, CB_{inst} or CB_{prop}) the context S1_1 and applies the corresponding bridge rules. As it happens for contexts, we can declare bridge rules lazily. Monotonicity guarantees that we cannot derive facts that should be retracted if some new bridge rule were added. Notice that bridge rules with side conditions (e.g. $\mathcal{R}_{up.}$, Bel-Clo) are implemented so that the applicability is automatically checked at run time.

The bridge rules of the MC system described in the previous section are defined on the basis of the map between formulas and their relative quotation mark names. This relation must be mechanized. This can be done in GETFOL by exploiting the *attachment* mechanism. Each individual constant symbol can be "attached" to a data structure, and each predicate and function symbol can be "attached" to a subroutine written in HGKM [Giunchiglia and Cimatti, 1991] (the implementation language of GETFOL). Attachments are declared by the commands battach and attach. Here are some examples:

```
GETFOL:: battach ~White1~ DAR ::WFF:White1;
GETFOL:: battach ~B1~   DAR ::PREDCONST:B1;
GETFOL:: attach makenot TO makenot;
GETFOL:: attach encode  TO encode
```

The first command attaches ~White1~ to the data structure representing the wff White1. The third command attaches the function symbols makeand to the HGKM function makeand, which maps two (data structure representing) wffs into (the data structure representing) their conjunction.

Following the attachments we can evaluate a term of the language. For instance, the result of the evaluation of

```
apply(~B1~,encode(makeand(~White1~,~White2~)))
```

is the wff B1(makeand(~White1~,~White2~)). Bridge rules are defined on the basis of evaluation. Indeed, dropping the quotation marks in the formal version of a bridge rule (e.g $\mathcal{R}_{dn.}$) corresponds to evaluating a term in its mechanized version (rdown). Analogously, putting the quotation marks around a wff in a formal version of a bridge rule (e.g. $\mathcal{R}_{up.}$) corresponds to building a term which evaluates to such a wff.

4 THE MECHANIZED SOLUTIONS

In this section we discuss the mechanization of inference using belief contexts. We show how the mechanized proofs of (some of) the puzzles formalized in [Cimatti and Serafini, 1995] and [Cimatti and Serafini, 1996] are obtained in GETFOL. For lack of space, we do not present the details of the formal solutions (see cited papers). We rather focus on the way proofs are built, and the conceptual structure of inference, to highlight the advantages of belief contexts.

The first puzzle we consider is the OTWM, i.e. the Original TWM puzzle stated in section 1. Axioms are declared in context S1 by means of the command axiom. The initial situation is formalized by stating that every agent can see its colleagues (e.g. A1seesA2), that this fact is a common belief (e.g. CB_A1seesA2), that king utterance ("At least one spot is white") is a common belief (ListenKing), and that all the spots are white (Spot).

```
GETFOL:: switchcontext S1;
GETFOL:: axiom A1seesA2: White2 imp B1("White2");
         ...
GETFOL:: axiom CB_A1seesA2:
   CB(makeimp("White2" apply("B1" encode("White2"))));
         ...
GETFOL:: axiom ListenKing:
   CB(makeor("White1" makeor("White2" "White3")));

GETFOL:: axiom Spot: White1 and White2 and White3;
```

The first situation describes the reasoning of the first wise man, which answers "I don't know". The conceptual structure of the proof is depicted by figure 3. First, we perform local reasoning in S1 by means of the LDL command. LDL selects atomic formulas and verifies their derivability from the facts of the context by applying a simple propositional decider. If a formula is derivable, then it is asserted as a fact (i.e. a labelled proof line) in the context, and printed out:

```
GETFOL:: LDL;
1   B3("White2")
2   B3("White1")
3   B2("White3")
4   B2("White1")
5   B1("White3")
6   B1("White2")
```

In this case, the asserted facts state that each wise men knows the color of the spot of his colleagues. We are interested in wise 1, who answers first. Therefore, we switch to the context S1_1, which is a model for wise 1, and we apply the rdownall command.

```
GETFOL:: switchcontext S1_1;
GETFOL:: rdownall;
1  White1 or (White2 or White3)
2  CB(makeor("White1",makeor("White2","White3")))
3  (not White2) imp B3(makenot("White2"))
4  CB(makeimp(makenot("White2"),
                  apply("B3",encode(makenot("White2")))))
5  White2 imp B3("White2")
6  CB(makeimp("White2",apply("B3",encode("White2"))))
7  (not White1) imp B3(makenot("White1"))
8  CB(makeimp(makenot("White1"),
                 apply("B3",encode(makenot("White1")))))
9  White1 imp B3("White1")
10 CB(makeimp("White1",apply("B3",encode("White1"))))
11 (not White3) imp B2(makenot("White3"))
12 CB(makeimp(makenot("White3"),
                  apply("B2",encode(makenot("White3")))))
13 White3 imp B2("White3")
14 CB(makeimp("White3",apply("B2",encode("White3"))))
15 (not White1) imp B2(makenot("White1"))
16 CB(makeimp(makenot("White1"),
                  apply("B2",encode(makenot("White1")))))
17 White1 imp B2("White1")
18 CB(makeimp("White1",apply("B2",encode("White1"))))
19 (not White3) imp B1(makenot("White3"))
20 CB(makeimp(makenot("White3"),
                  apply("B1",encode(makenot("White3")))))
21 White3 imp B1("White3")
22 CB(makeimp("White3",apply("B1",encode("White3"))))
23 (not White2) imp B1(makenot("White2"))
24 CB(makeimp(makenot("White2"),
                  apply("B1",encode(makenot("White2")))))
25 White2 imp B1("White2")
26 CB(makeimp("White2",apply("B1",encode("White2"))))
27 White2
28 White3
```

The labelling of facts is local to a context, i.e. facts of different contexts can have the same label. `rdownall` can be interpreted as a focusing operation: it allows us to represent information *about* the agent as information *in the model of* the agent. Notice that facts 5 and 6 in context S1 are converted into facts 27 and 28 with a simpler format where the belief predicate has been eliminated. The other facts are common belief, and can be propagated as they are at any level of depth.

After the execution of `rdownall`, the model of wise 1 in the first situation contains a reasonable amount of knowledge. Then, we switch back to S1 and we apply the command `belclo_a1`, which mechanizes the belief closure bridge rule:

```
GETFOL:: switchcontext S1;
GETFOL:: belclo_a1 White1;
7   ARF1(GAMMA28,~White1~)      (7)
8   not B1(~White1~)      (7)
```

The mechanization of belief closure checks with a propositional decider whether the formula White1 follows from the theorems of context S1_1. As it does not, then it automatically generates the relevance assumption 7, where GAMMA28 is a newly declared individual constant representing the relevant facts of S1_1 that have been taken into account. Then, it asserts the ignorance statement. Notice that deduction in GETFOL is implemented in a natural-deduction style, and assumptions are kept distinct from axioms. Both fact 7 and fact 8 depend on fact 7: this is stated by the dependency list (7) printed on the right of both facts.

The second situation is described by the tree of contexts with root S2. The situation is very much as the first, with the addition of an axiom corresponding to the answer of the first wise man. The proof for the second situation, where wise 2 answers "I don't know", is obtained with the same commands of the first. The schema of the proof for the third situation, where the third wise answers "My spot is white", is depicted in figure 4. The additional axioms with respect to the first situation are:

```
GETFOL:: switchcontext S3;
GETFOL:: axiom ....
GETFOL:: axiom Answer1:
   CB(makenot(apply(~B1~,encode(~White1~)))) ;
GETFOL:: axiom Answer2:
   CB(makenot(apply(~B2~,encode(~White2~)))) ;
```

Answer1 and Answer2 state that it is a common belief that wise 1 and 2 don't know that their spot is white. After performing local reasoning with LDL, we switch to context S3_3 and we apply rdownall.

```
GETFOL:: LDL;
        ....
GETFOL:: switchcontext S3_3;
GETFOL:: rdownall;
1   not B1(~White1~)
    ....
3   not B2(~White2~)
    ....
```

With respect to the first situation, we obtain the additional facts 1 and 3, which derive from the axioms Answer1 and Answer2. In order to solve the puzzle, wise 3 reasons by contradiction, assuming that his spot is black (fact 33).

```
GETFOL:: assume not White3;
33   not White3      (33)
GETFOL:: LDL;
       ....
36   B2(~White1~)
37   B2(makenot(~White3~))      (33)
       ....
```

Then, he reasons about the beliefs of wise 2 under assumption 33.

```
GETFOL:: switchcontext S3_32;
GETFOL:: rdownall ;
1   not White3
2   White1
       ....
```

Notice that in context S3_32 fact 1 does not show any dependency on assumption 33, although this dependency is part of the logic (and is stored in the system too). This allows for a local view of the context, i.e. the user (and the deductive routines) is not forced to take facts of other contexts. In context S3_32 an assumption is performed (labelled 33 by chance) to formalize that wise 2 hypothesizes that his spot is black.

```
GETFOL:: assume not White2;
33   not White2      (33)
GETFOL:: LDL;
       ....
38   B1(makenot(~White3~))
39   B1(makenot(~White2~))      (33)
GETFOL:: switchcontext S3_321;
GETFOL:: rdownall ;
1   not White2
2   not White3
       ....
7   White1 or (White2 or White3)
       ....
GETFOL:: LDL;
       ....
36   White1
```

In context S3_321, via local reasoning, we derive fact 36. This is the conclusion we need in order to refute the assumptions our reasoning is based on. We switch back to context S3_32, and we reflect up with command rup fact 36 (the notation #(White1) stands for the fact whose wff is White1).

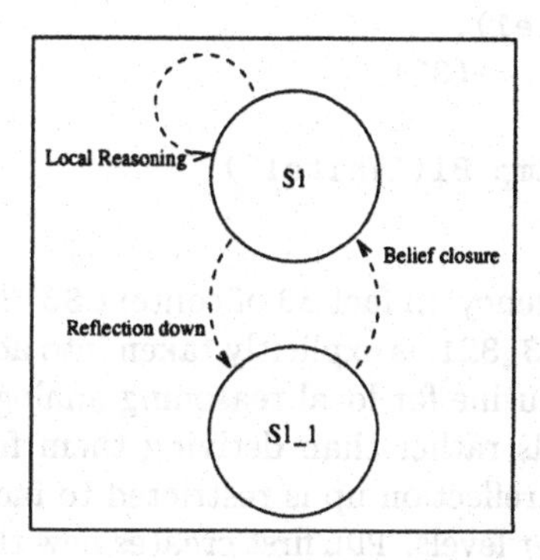

Figure 3. First situation for the OTWM

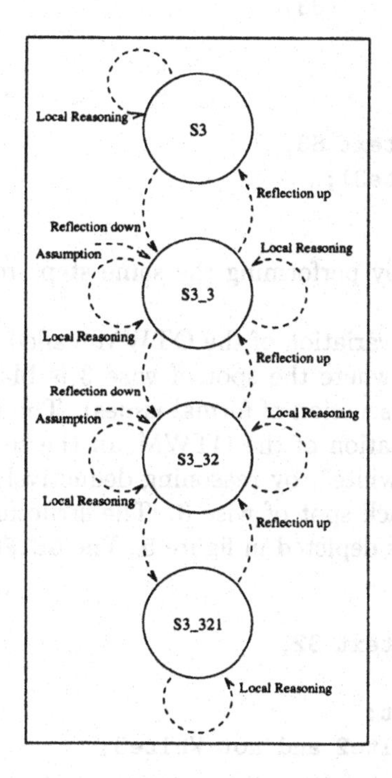

Figure 4. Third situation for the OTWM

```
GETFOL:: switchcontext S3_32;
GETFOL:: rup #(White1);
40   B1(~White1~)      (33)
GETFOL:: PDL;
41   (not White2) imp B1(~White1~)
42   White2
```

Notice that the dependency on fact 33 of context S3_32, hidden when showing fact 36 of context S3_321, is explicitly taken into account when reflecting up fact 40. PDL is a routine for local reasoning analogous to LDL. However, it tries to prove literals rather than deriving them from assumptions, the motivation being that reflection up is restricted to facts depending at most on assumption of higher levels. PDL first creates new theorems starting from facts with dependencies by applying imply introduction (e.g. theorem 41 in context S3_32); then it applies LDL to the theorems of the context.

```
GETFOL:: switchcontext S3_3;
GETFOL:: rup #(White2);
40   B2(~White2~)      (33)
GETFOL:: PDL ;
     ....
43   White3
GETFOL:: switchcontext S3;
GETFOL:: rup #(White3);
7   B3(~White3~)
```

The proof concludes by performing the same steps reflecting up to derive B3(~White3~).

We consider now a variation of the OTWM, called WWBTWM (White-White-Black TWM), where the spot of wise 3 is black (see [Cimatti and Serafini, 1996] for a discussion of formal issues). The first situation is analogous to the first situation of the OTWM. In the second situation wise 2 answers "My spot is white", by reasoning deductively on the ignorance of wise 1 and on the black spot of wise 3. The structure of the inference in the second situation is depicted in figure 5. The GETFOL commands are the following:

```
GETFOL:: switchcontext S2;
GETFOL:: axiom ....
GETFOL:: axiom Spot:
     White1 and White2 and not White3;
GETFOL:: LDL;
GETFOL:: switchcontext S2_2;
GETFOL:: rdownall ;
GETFOL:: assume not White2;
```

```
GETFOL:: LDL;
GETFOL:: switchcontext S2_21;
GETFOL:: rdownall ;
GETFOL:: LDL ;
GETFOL:: switchcontext S2_2;
GETFOL:: rup #(White1);
GETFOL:: PDL;
GETFOL:: switchcontext S2;
GETFOL:: rup #(White2);
7   B2(~White2~)
```

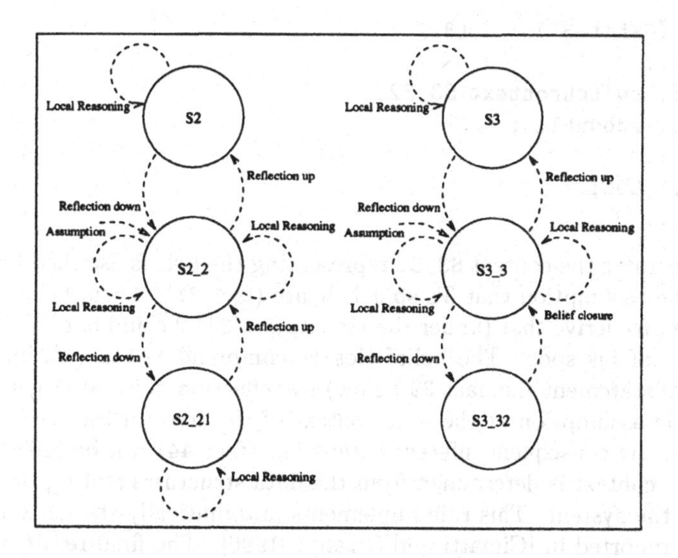

Figure 5. Second and third situations for the WWBTWM

An analysis of the similarity between this proof and the third proof in the OTWM can be found in [Cimatti and Serafini, 1996] (intuitively, wise 2 in this case performs the very same reasoning that wise 3 ascribes to him in the OTWM under the assumption `not White3`). Here we stress the similarity between the sequence of commands: compare the trace of wise 3 in the OTWM, and the following trace of wise 2 in the WWBTWM. Also, compare the conceptual structure of the above deduction to the proof of the third situation in the OTWM (figures 4 and 5).

In the third situation, wise 3 answers "My spot is black". The structure of the proof is shown in figure 5. The "downward" part of the proof is obtained by repeated applications of `LDL`, `rdownall` and assumption generation.

```
GETFOL:: switchcontext S3;
```

```
GETFOL:: axiom ....
GETFOL:: LDL ;
GETFOL:: switchcontext S3_3;
GETFOL:: rdownall ;
      ....
26   B2(~White2~)
      ....
GETFOL:: assume White3;
31   White3     (31)
GETFOL:: LDL;
      ....
33   B1(~White3~)     (31)
      ....
GETFOL:: switchcontext S3_32;
GETFOL:: rdownall ;
      ....
GETFOL:: LDL;
      ....
```

This generates the context S3_32 representing the beliefs ascribed by 3 to 2 under the assumption that 3's spot is white (fact 31). Then, belief closure is applied to derive that (under the assumption 31) 2 could not have known the color of his spot. The belief closure command **belclo_a2** infers the relevance statement (i.e. fact 39 below) by reflection down after generating a suitable assumption at the root context of the tree (in this case S3). In this way, the consequent inference steps (e.g. fact 44) can be reflected up. The root context is determined from the data structures storing the bridge rules in the system. This rule implements automatically the derived proof pattern reported in [Cimatti and Serafini, 1996]. The final result depends on the assumption (embedded in the puzzle, and expressed by fact 7) that wise 3 believes that facts from 1 to 34 (of context S3_32) are the only facts available to wise 2 in order to infer that his spot is white.

```
GETFOL:: switchcontext S3_3;
GETFOL:: belclo_a2 White2;
Switching to context S3
7   B3(apply2(~ARF2~,~GAMMA34~,
              encode(~White2~)))  (7)
Switching to context S3_3
38   ARF2(GAMMA34,~White2~)
39   not B2(~White2~)      (31)
GETFOL:: taut FALSE by
     #(B2(~White2~)), #(not B2(~White2~));
40   FALSE     (31)
GETFOL:: noti #(FALSE) White3;
```

```
44   not White3

GETFOL:: switchcontext S3;
GETFOL:: rup #(not White3);
8   B3(mknot("White3"))     (7)
```

Some remarks are in order. First, the proofs shown above have a standard form (see figures 3, 4 and 5): there is a "downward" part, and an "upward" part, obtained by iterating reflection down, local reasoning and reflection up. This structure is a standard pattern in reasoning about propositional attitudes (see for instance [Haas, 1986; Konolige, 1986]). The use of belief contexts allows us to separate knowledge in a modular way: the structure of the formal system makes it clear what information has to be taken into account in local reasoning. It is interesting to notice that this has a formal counterpart in the proof theory of ML systems, e.g. sublevel property theorem, normal form theorem [Giunchiglia and Serafini, 1993].

Second, there are only few conceptual operations: reflection down (implemented by the command `rdownall`), hypothesis generation (the command `assume`), local reasoning (e.g. `LDL`, `PDL`) reflection up, and belief closure. This suggests that general proof strategies can be defined as simple tactic-like primitives (however, this is an open issue).

Third, a substantial part of reasoning is simple propositional reasoning local to a context. In the case of our examples, a propositional decider has been used in a very simple way, i.e. applying it to the goal formula and the conjunction of all facts of the context. Nevertheless, the intrinsic structure of the proof allows to limit the formulas to consider, therefore there is no problem of efficiency. This gives implementational advantages with respect to inference in an unstructured, flat (e.g. modal) logic. For the same reason, the results of customized deciders exploiting the structure of contexts (which are currently under test) appear to be extremely promising.

5 CONCLUSIONS

Belief contexts are an expressive and modular framework for the formalization of propositional attitudes in a multiagent environment. In this paper we have show how they provide for many advantages from the implementational point of view. First, we have shown how reasoning about mutual and nested beliefs, common belief, ignorance and ignorance ascription, can be mechanized using belief contexts in a very general and structured way. Then we have shown how very simple and meaningful inference steps can be combined in a uniform framework and used to solve several versions of the TWM.

ITC-IRST, Trento, Italy.

REFERENCES

[Cimatti and Serafini, 1995] A. Cimatti and L. Serafini. Multi-Agent Reasoning with Belief Contexts: the Approach and a Case Study. In M. Wooldridge and N. R. Jennings, editors, *Intelligent Agents: Proceedings of 1994 Workshop on Agent Theories, Architectures, and Languages*, number 890 in Lecture Notes in Computer Science, pages 71–85. Springer Verlag, 1995. Also IRST-Technical Report 9312-01, IRST, Trento, Italy.

[Cimatti and Serafini, 1996] A. Cimatti and L. Serafini. Multi-Agent Reasoning with Belief Contexts II: Elaboration Tolerance. In *Proc. 1st Int. Conference on Multi-Agent Systems (ICMAS-95)*, pages 57–64, 1996. Also IRST-Technical Report 9412-09, IRST, Trento, Italy. Presented at *Commonsense-96*, Third Symposium on Logical Formalizations of Commonsense Reasoning, Stanford University, 1996.

[Creary, 1979] L. G. Creary. Propositional Attitudes: Fregean representation and simulative reasoning. In *Proc. of the 6th International Joint Conference on Artificial Intelligence*, pages 176–181, 1979.

[Criscuolo *et al.*, 1994] G. Criscuolo, F. Giunchiglia, and L. Serafini. A Foundation of Metalogical Reasoning: OM pairs (Propositional Case). Technical Report 9403-02, IRST, Trento, Italy, 1994.

[Giunchiglia and Cimatti, 1991] F. Giunchiglia and A. Cimatti. HGKM User Manual - HGKM version 2. Technical Report 91-0009, DIST - University of Genova, Genova, Italy, 1991.

[Giunchiglia and Serafini, 1991] F. Giunchiglia and L. Serafini. Multilanguage first order theories of propositional attitudes. In *Proceedings 3rd Scandinavian Conference on Artificial Intelligence*, pages 228–240, Roskilde University, Denmark, 1991. IOS Press. Also IRST-Technical Report 9001-02, IRST, Trento, Italy.

[Giunchiglia and Serafini, 1993] F. Giunchiglia and L. Serafini. On the Proof Theory of Hierarchical Meta-Logics. Technical Report 9301-07, IRST, Trento, Italy, 1993.

[Giunchiglia and Serafini, 1994] F. Giunchiglia and L. Serafini. Multilanguage hierarchical logics (or: how we can do without modal logics). *Artificial Intelligence*, 65:29–70, 1994. Also IRST-Technical Report 9110-07, IRST, Trento, Italy.

[Giunchiglia and Traverso, 1996] F. Giunchiglia and P. Traverso. A Metatheory of a Mechanized Object Theory. *Artificial Intelligence*, 80(2):197–241, 1996. Also IRST-Technical Report 9211-24, IRST, Trento, Italy, 1992.

[Giunchiglia and Weyhrauch, 1991] F. Giunchiglia and R.W. Weyhrauch. FOL User Manual - FOL version 2. Manual 9109-08, IRST, Trento, Italy, 1991. Also DIST Technical Report 91-0006, University of Genova.

[Giunchiglia *et al.*, 1993] F. Giunchiglia, L. Serafini, E. Giunchiglia, and M. Frixione. Non-Omniscient Belief as Context-Based Reasoning. In *Proc. of the 13th International Joint Conference on Artificial Intelligence*, pages 548–554, Chambery, France, 1993. Also IRST-Technical Report 9206-03, IRST, Trento, Italy.

[Giunchiglia, 1993] F. Giunchiglia. Contextual reasoning. *Epistemologia, special issue on I Linguaggi e le Macchine*, XVI:345–364, 1993. Short version in Proceedings IJCAI'93 Workshop on Using Knowledge in its Context, Chambery, France, 1993, pp. 39–49. Also IRST-Technical Report 9211-20, IRST, Trento, Italy.

[Giunchiglia, 1994] F. Giunchiglia. GETFOL Manual - GETFOL version 2.0. Technical Report 92-0010, DIST - University of Genoa, Genoa, Italy, March 1994.

[Haas, 1986] A. R. Haas. A Syntactic Theory of Belief and Action. *Artificial Intelligence*, 28:245–292, 1986.

[Hobbs and Moore, 1985] J.R. Hobbs and R.C. Moore, editors. *Formal Theories of Commonsense World*. Ablex Publishing Corporation, Norwood, New Jersey, 1985.

[Konolige, 1986] K. Konolige. *A deduction model of belief*. Pitman, London, 1986.

[McCarthy, 1990] J. McCarthy. Formalization of Two Puzzles Involving Knowledge. In V. Lifschitz, editor, *Formalizing Common Sense - Papers by John McCarthy*, pages 158–166. Ablex Publishing Corporation, 1990.

[Moore, 1982] R.C. Moore. The role of logic in knowledge representation and commonsense reasoning. In *National Conference on Artificial Intelligence*. AAAI, 1982.

[Prawitz, 1965] D. Prawitz. *Natural Deduction - A proof theoretical study.* Almquist and Wiksell, Stockholm, 1965.

[Tarski, 1956] A. Tarski. *Logic, Semantics, Metamathematics.* Oxford University Press, 1956.

[Weyhrauch, 1980] R.W. Weyhrauch. Prolegomena to a Theory of Mechanized Formal Reasoning. *Artificial Intelligence*, 13(1):133–176, 1980.

PAUL PIWEK AND EMIEL KRAHMER

PRESUPPOSITIONS IN CONTEXT: CONSTRUCTING BRIDGES

1 INTRODUCTION

Traditionally, a distinction is made between that what is asserted by uttering a sentence and that what is presupposed. Presuppositions are characterized as those propositions which persist even if the sentence which triggers them is negated. Thus 'The king of France is bald' presupposes that *there is a king of France*, since this follows from both 'The king of France is bald' and 'It is not the case that the king of France is bald'.

Stalnaker [1974] put forward the idea that a presupposition of an asserted sentence is a piece of information which is assumed by the speaker to be part of the common background of the speaker and interpreter. The *presuppositions as anaphors* theory of Van der Sandt [1992]—currently the best theory of presupposition as far as empirical predictions are concerned [Beaver, 1997, p. 983]—can be seen as one advanced realization of Stalnaker's basic idea. The main insight of Van der Sandt is that there is an interesting correspondence between the behaviour of anaphoric pronouns in discourse and the projection of presuppositions (i.e. whether and how presuppositions survive in complex sentences). Like most research in this area, Van der Sandt's work concentrates on the interaction between presuppositions and the *linguistic* context (i.e. the preceding sentences). However, not only linguistic context interacts with presuppositions. Consider:

(1) a. If John buys a car, he checks the motor first.
    b. John walked into the room. The chandelier sparkled brightly.
    c. Mary traded her old car in for a new one. The motor was broken.

All three examples are instances of the notorious *bridging* phenomenon [Clark, 1975]. Example (1.a) contains a definite description, *the motor*, which triggers the presupposition that there is a motor. Intuitively, (1.a) as a whole does not presuppose the existence of a motor; this presupposition is 'absorbed' by the antecedent. However, because there is no proper antecedent for this definite description, the theory of Van der Sandt [1992] predicts that the presupposition that there is a motor is *accommodated* (where accommodation—the term is due to Lewis [1979]—amounts to simply adding the presupposition to the context). This fails to do justice to the intuition that the mentioning of a car somehow licenses the use of *the motor* and that the motor is part of the car which John buys. Thus, for the correct treatment of this example, a rather trivial piece of world knowledge

P. Bonzon, M. Cavalcanti and R. Nossum (eds.), Formal Aspects of Context, 85–106.

is needed: cars have motors. In the presence of such background knowledge an interpreter will be able to construct a bridge between the would-be antecedent (*a car*) and the presupposition/anaphor (*the motor*). Example (1.b) can be explained along similar lines; the interpreter has to construct a bridge between *a room* and *the chandelier*. Unfortunately, things are a bit more complicated for this example. After all, the interpreter will not be able to use background knowledge such as rooms have chandeliers, since there are many chandelier-less rooms. Example (1.c) illustrates yet another complication: granted that cars have motors, with which of the two cars introduced in the first sentence of example (1.c) should *the motor* from the second sentence be associated?

For all these examples, the theory from Van der Sandt [1992] predicts that the presuppositions are accommodated, due to the fact that the non-linguistic context is not taken into account. In this article, we want to get a formal grip on the way in which context influences the behaviour of presuppositions. Before we describe how we intend to do this, let us first describe the notion of context we are interested in. There are various uses of the term 'context'. Bunt [1995] characterizes context as all those factors which are relevant to the understanding of communicative behaviour, and he goes on to distinguish five major dimensions: the linguistic context, the semantic context, the physical context, the social context and the cognitive context. For presuppositions in general, and for bridging in particular, the following seem most relevant: the *linguistic* context, as this will contain the antecedents from which a bridge has to be constructed, and the *cognitive* context, which according to Bunt includes the attentional state and the world knowledge of an interlocutor. Throughout this article we will therefore focus on the linguistic and the cognitive context.

The resulting global picture is as follows: an interpreter tries to understand a sentence in some context Γ. This context contains representations of the preceding discourse (the linguistic context) as well as background knowledge (the cognitive context). The interlocutor assumes that parts of her context are, to some extent, public. That is, they form what the interlocutor assumes to be the common ground. In this article, we are particularly interested in how interlocutors use the context to come to an understanding of the current sentence, and how they can adjust their context on the fly, so to speak, when the current sentence calls for such an adjustment. This brings out the extreme flexibility of context in natural language communication: speaker and hearer constantly attempt to align their representations. For more details on the role of context in communication, we refer to Piwek [1998].

The claim that context, and more specifically world knowledge, has an influence on presupposition projection is hardly revolutionary, the question is how to *account* for this influence. We argue that employing a class of mathematical formalisms known as *Constructive Type Theories* (CTT, see e.g. [Martin-Löf, 1984; Barendregt, 1992]) allows us to answer this question. To do so, we reformulate Van der Sandt's theory in terms of CTT. CTT differs from other proof systems in that for each proposition which is proven, CTT also delivers a proof object which shows how the proposition was proven. As we shall see, the presence of these proof objects is useful from the presuppositional point of view. Additionally, CTT contexts contain *more* information than is conveyed by the ongoing discourse, and there is a formal interaction between this 'background knowledge' and the representation of the current discourse. This means that the reformulation of Van der Sandt's theory in terms of CTT is not just a nice technical exercise, but actually creates some interesting new possibilities where the interaction between presupposition resolution and world knowledge context is concerned. To illustrate this, we show that the resulting system facilitates the treatment of the notorious bridging phenomenon illustrated above. In particular, it will be shown that many of the observations made in Clark's [1975] seminal 'bridging' paper have nice CTT counterparts. We propose to come to a so-called *determinate* bridge by imposing two conditions: *effort* and *plausibility*. Finally, we discuss an explorative study we conducted to find support for our analysis. In this article we will not dig too deep into the formalities of our approach of 'presupposition projection as proof construction', for that we refer to [Krahmer and Piwek, to appear].

2 PRESUPPOSITIONS AS ANAPHORS

Van der Sandt [1992] proposes to *resolve* presuppositions, just like anaphoric pronouns are resolved in *Discourse Representation Theory* (DRT, [Kamp and Reyle, 1993]. In DRT, linguistic contexts are modelled as *Discourse Representation Structures* (DRSs). A DRS consists of a set of *discourse referents* and a set of *conditions* on these referents. The discourse referents can be seen as representatives for the objects which are *introduced* in the discourse, and the conditions can be seen as assignments of properties to these objects. To resolve presuppositions in DRT, Van der Sandt [1992] develops a meta-level resolution algorithm. The input of this algorithm is an underspecified DRS, which contains one or more unresolved presuppositions. When all these presuppositions have been resolved, a proper DRS remains, which can be interpreted in the standard way.[1] Let us consider the following example, and its Van der Sandtian representation:

(2) If John buys a pantechnicon, he'll adore the vehicle.

[1] In [Krahmer, 1998], Van der Sandt's theory is combined with a version of DRT with a partial interpretation. In this way, DRSs which contain unresolved presuppositions can also be interpreted. It is shown that this has several advantages.

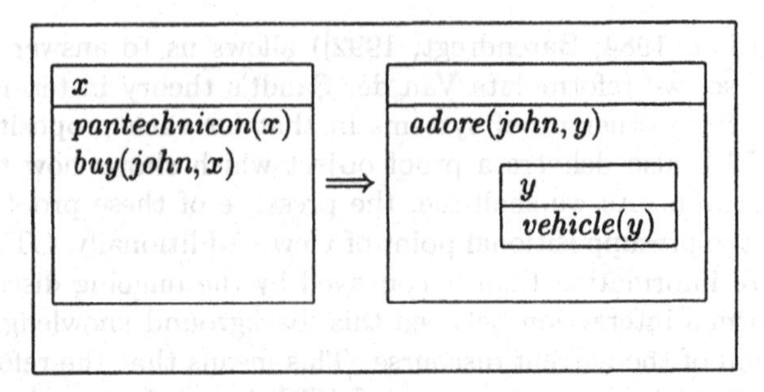

This DRS consists of a complex condition, containing two sub-DRSs, one for the antecedent and one for the consequent of (2). The antecedent DRS introduces a referent x. This x stands for a pantechnicon which is bought by John (where 'John' is represented by a constant (*john*) for the sake of simplicity). The definite description *the vehicle* presupposes the existence of a vehicle. This is modelled by adding an embedded, presuppositional DRS to the consequent DRS introducing a referent y which is a vehicle. The consequent DRS additionally contains the condition that this presupposed vehicle is adored by John. To *resolve* the presuppositional DRS, we do what we would do to resolve a pronoun: look for a suitable, accessible antecedent. In this case, we find one: the discourse referent x introduced in the antecedent *is* accessible[2] and suitable since a pantechnicon (i.e. a removal truck) is a vehicle. Exactly *how* this information can be employed in Van der Sandt's theory is not obvious. For now, we will simply assume that we can *bind* the presupposition, which results in the following DRS, which can be paraphrased as 'if John buys a pantechnicon, he'll adore it'.[3]

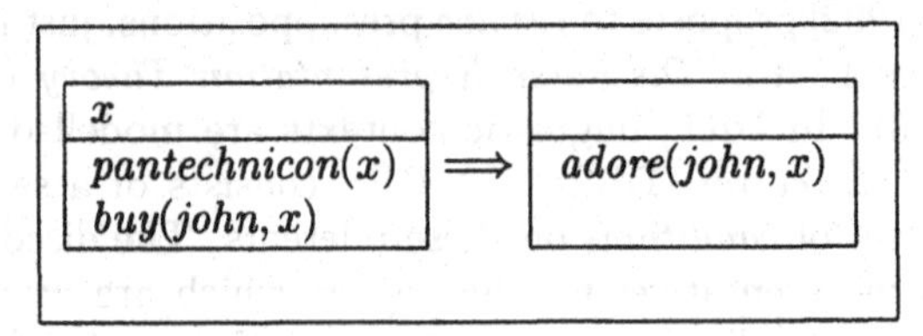

In principle, anaphoric pronouns are always bound. For presuppositions this is different: they can also be accommodated, provided the presupposition contains sufficient descriptive content. Reconsider example (2) again: on Van der Sandt's approach (globally) *accommodating* the presupposition associated with *the vehicle* amounts to removing the presuppositional DRS from the consequent DRS and placing it in the main DRS, which would result in the following DRS.

[2]In DRT, the generalization is that discourse referents introduced in an antecedent DRS are accessible from the consequent DRS.

[3]This DRS (as the previous one) is presented in the usual 'pictorial' fashion. Elsewhere in this paper we also use a linear notation which we trust to be self-explanatory. E.g. in this linear notation the current DRS looks as follows: $[\,|\,[x \mid pantechnicon(x), buy(john, x)] \Longrightarrow [\,|\, adore(john, x)]]$.

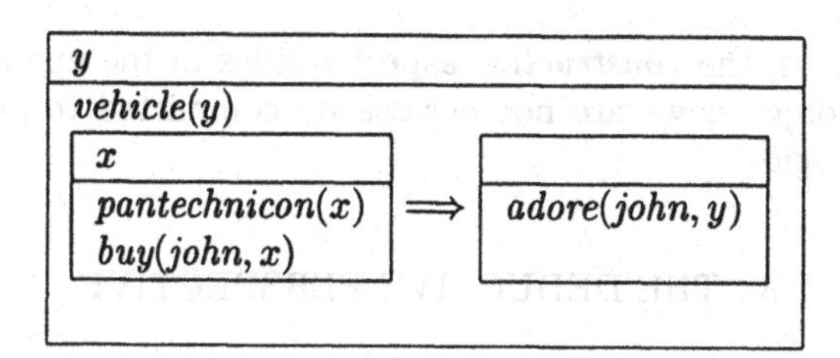

This DRS represents the 'presuppositional' reading of (2), which may be paraphrased as 'there is a vehicle and if John buys a pantechnicon, he'll adore the aforementioned vehicle'. Now we have *two* ways of dealing with the presupposition in example (2), so the question may arise which of these two is the 'best' one. To answer that question, Van der Sandt defines some general rules for preferences, which may be put informally as follows: 1. Binding is preferred to accommodation, 2. Binding is preferred as low as possible, and 3. Accommodation is preferred as high as possible (thus, preferably in the main DRS). The third preference rule seems to suggest that there is more than one way to accommodate a presupposition, and indeed there is. To illustrate this, consider:

(3) It is not true that I adore John's pantechnicon, since he doesn't have one!

Here, the definite NP *John's pantechnicon* presupposes that John has a pantechnicon. If we globally accommodate this presupposition (that is, the presupposition 'escapes' from the scope of the negation and is placed in the main DRS), we would end up with an inconsistent DRS, expressing that John has a pantechnicon, which is contradicted by the *since*-clause. Van der Sandt [1992, p. 367] defines a number of conditions on accommodation, of which consistency is one. Since in the case of (3) global accommodation yields an inconsistent DRS, *local accommodation* of the presupposition is preferred, where local means within the scope of the negation. The result can be paraphrased as 'it is not true that John has a pantechnicon and that I adore it, since he doesn't have one'.

In the next section, we discuss CTT and show how Van der Sandt's approach can be rephrased in terms of it. In the section thereafter, we will see how the examples in (1), which are problematic for Van der Sandt's approach as it stands, can be dealt with. We believe that the CTT approach leads to better results than adding a proof system to DRT, as done in e.g. [Saurer, 1993]. The main advantage of CTT is that it is a standard proof system developed in mathematics with well-understood meta-theoretical properties (see [Ahn and Kolb, 1990] for discussion on the advantages of reformulating DRT in CTT). Moreover, the presence of explicit proof objects in CTT turns out to have some additional advantages for our present purposes. For us, the *constructive* aspect resides in the explicit construction of proof-objects; we are not necessarily committed to an underlying intuitionistic logic.

3 THE DEDUCTIVE PERSPECTIVE

The deductive approach to discourse

We introduce CTT by comparing it with DRT; this comparison is based on Ahn and Kolb [1990], who present a formal translation of DRSs into CTT expressions. A context in CTT is modelled as a sequence of introductions. Introductions are of the form $V : T$, where V is a variable and T is the type of the variable. Consider example (4.a) and its DRT representation (4.b) (in the linear notation, cf. footnote 3).

(4) a. John drives a vehicle.
 b. $[x|\textit{vehicle}(x), \textit{drives}(\textit{john}, x)]$

A discourse referent can be modelled in CTT as a variable. A referent is added to the context by means of an introduction which not only adds the variable but also fixes its type. We choose *entity* as the type of discourse referents. The type *entity* itself also requires introduction. Since entity is a type, we write: *entity*:*type*.

The type *entity* should only be used in the introduction x:*entity* if *entity*:*type* is already part of the context. This way, one introduction depends on another introduction, hence a context is an ordered sequence of introductions. The type *type* also requires introduction. The introduction is, however, not carried out in the context; it is taken care of by an axiom which says that *type*:□ (where □ is to be understood as the 'mother of all types') can be derived in the empty context ($\varepsilon \vdash \textit{type} : \square$).

DRT's conditions correspond to introductions $V : T$, where T is of the type *prop* (short for proposition, which comes with the following axiom: $\varepsilon \vdash \textit{prop} : \square$). For instance, the introduction $y : (\textit{vehicle} \cdot x)$ corresponds to the condition *vehicle*(x). The type *vehicle* $\cdot x$ (of type *prop*) is obtained by applying the type *vehicle* to the object x. Therefore, it depends on the introductions of x and *vehicle*. Since *vehicle* $\cdot\, x$ should be of the type *prop*, vehicle must be a (function) type from the set of entities into propositions, i.e. $\textit{vehicle} : \textit{entity} \rightarrow \textit{prop}$.

The introduction $y : (\textit{vehicle} \cdot x)$ involves the variable y (of the type $\textit{vehicle} \cdot x$). The variable y is said to be an inhabitant of *vehicle*$\cdot x$. Curry and Feys [1958] came up with the idea that propositions can be seen as classifying proofs (this is known as the 'propositions as types—proofs as objects' interpretation). This means that the aforementioned introduction states that there is a proof y for the proposition *vehicle* $\cdot x$. The second DRS condition (*drive(john, x)*) can be dealt with along the same lines. Assume that *drive* is a predicate which requires two arguments of the type *entity*, this yields $z : drive \cdot x \cdot john$. (The '$\cdot$' (representing function application) is left-associative, thus $f \cdot x \cdot y$ should be read as $(f \cdot x) \cdot y$). In sum, the CTT counterpart to the DRS (4.b) consists of the following three introductions: $x : entity, y : vehicle \cdot x, z : drive \cdot x \cdot john$.

Dependent Function Types

In DRT, the proposition *Everything moves* is translated into the implicative condition $[x \mid thing(x)] \Longrightarrow [\ \mid move(x)]$. In CTT, this proposition corresponds to the type $(\Pi x : entity.move \cdot x)$, which is a dependent function type. It describes functions from the type *entity* into the type $move \cdot x$. The range of such a function $(move \cdot x)$ depends on the object x to which it is applied. Suppose that we have an inhabitant f of this function type, i.e. $f : (\Pi x : entity.move \cdot x)$. Then we have a function which, when it is applied to an arbitrary object y, yields an inhabitant of the proposition $move \cdot y$. Thus, f is a constructive proof for the proposition that *Everything moves*.

Of course, function types can be nested. Consider the predicate *drive*. Above we suggested to introduce it as a function from entities ('the driver') to entities ('the thing being driving') to propositions. One could, however, argue that the second argument of *drive* ('the thing being driven') can only be a vehicle. In that case, *drive* would have to be introduced as function from entities to entities to another function (i.e. the function from a proof that the second entity is a vehicle to a proposition), that is $drive : (\Pi y : entity.(\Pi x : entity.(\Pi p : vehicle \cdot x.prop)))$. We will abbreviate this as $drive : ([y : entity, x : entity, p : vehicle \cdot x] \Rightarrow prop)$.

Deduction

The core of CTT consists of a set of derivation rules with which one can determine the type of an object in a given context. These rules are also suited for searching for an object belonging to a particular type. There is, for instance, a rule which is similar to modus ponens in classical logic (in the rule below, $T[x := a]$ stands for a T such that all free occurrences of x in T have been substituted by a. Furthermore, $\Gamma \vdash E : T$ means that in context Γ, the statement $E : T$ is provable):

$$\frac{\Gamma \vdash F : (\Pi x : A.B) \qquad \Gamma \vdash a : A}{\Gamma \vdash F \cdot a : B[x := a]}$$

For instance, if a context Γ contains the introduction $b{:}entity$ as well as the introduction $g : (\Pi y : entity.move \cdot y)$ ('everything moves'), then we can use this rule to find an inhabitant of the type $move \cdot b$. In other words, our goal is to find a substitution S such that $\Gamma \vdash P : move \cdot b[S]$. The substitution S

should assign a value to P. P is a so-called *gap*.[4] A CTT expression with a gap is an underspecified representation of a proper CTT expression: if the gap is filled, then a proper CTT expression is obtained. The deduction rule tells us that $(g \cdot b)$ can be substituted for P, if $\Gamma \vdash g : (\Pi y : \mathit{entity}.\mathit{move} \cdot y)$ and if $\Gamma \vdash b : \mathit{entity}$. Both so-called *judgements* are valid, because we assumed that $g : (\Pi y : \mathit{entity}.\mathit{move} \cdot y)$ and $b : \mathit{entity}$ are members of Γ. Thus, we can conclude that $\Gamma \vdash (g \cdot b) : \mathit{move} \cdot b$.

Presuppositions as Gaps

Van der Sandt's presuppositional DRSs can be seen as a kind of 'proto DRSs' for which the presuppositional representations have not yet been resolved. Only after resolution and/or accommodation of the presuppositions is a proper DRS produced. Analogously, in CTT terms, a construction algorithm could translate a sentence into a proto type before a proper type (of the type *prop*) is returned. This proper type (i.e. proposition) can then be added to the main context by introducing a fresh proof for it. Let us reconsider example (2), repeated below as (5), together with the appropriate proto type for this sentence in (6).

(5) If John buys a pantechnicon, he'll adore the vehicle.
(6) $[x : \mathit{entity}, y : \mathit{pantechnicon} \cdot x, z : \mathit{buy} \cdot x \cdot \mathit{john}] \Rightarrow$
$$(\mathit{adore} \cdot Y \cdot \mathit{john})_{[Y : \mathit{entity}, P : \mathit{vehicle} \cdot Y]}$$

In words: if x is an entity and y a proof that x is a pantechnicon and z is a proof that it is bought by John, then there exists a proof that John adores Y, where Y is a gap to be filled by an entity for which we can prove that it is a vehicle. The (subscripted) presuppositional annotation consists of a sequence of introductions with gaps.

Filling the Gaps: Binding v. Accommodation

Suppose we want to evaluate the CTT representation (6) given some context Γ. Before we can do that we have to resolve the presupposition by filling the gap. For this purpose, we have developed an algorithm which operates on proto-types and CTT contexts, based on Van der Sandt's presupposition resolution algorithm (see [Krahmer and Piwek, to appear] for technical details). The first thing we do after starting the resolution process, is try to 'bind' the presuppositional gap. The question whether we can bind the presupposition triggered by the vehicle in example (5) can be phrased in CTT as follows: is there a substitution S such that the following can be proven?

[4]In Piwek [1997; 1998] it is shown how these same gaps can be used in the analysis of questions. Piwek argues that questions introduce gaps, which can be filled by extending the context of interpretation with the answer provided by the dialogue participant. A question is answered, when the associated gaps can be filled.

$$(7)\ \Gamma, x : \mathit{entity}, y : \mathit{pantechnicon} \cdot x, z : \mathit{buy} \cdot x \cdot \mathit{john} \vdash (Y : \mathit{entity}, P : \mathit{vehicle} \cdot Y)[S]$$

In words: is it possible to prove the existence of a vehicle from the global context Γ extended with the local context (the antecedent)? The answer is: that depends on Γ. Suppose for the sake of argument that Γ itself does not contain any vehicles, but that it does contain the information that a pantechnicon is a vehicle. Technically, this means that the following function is a member of Γ:

$$(8)\ f : ([a : \mathit{entity}, b : \mathit{pantechnicon} \cdot a] \Rightarrow (\mathit{vehicle} \cdot a))$$

Given this function, we find a substitution S for (7), mapping Y to x and P to $(f \cdot x \cdot y)$ (which is the result of applying the aforementioned function f to x and y). So we fill the gaps using the substitution S, remove the annotations (which have done their job) and continue with the result:

$$(9)\ [x : \mathit{entity}, y : \mathit{pantechnicon} \cdot x, z : \mathit{buy} \cdot x \cdot \mathit{john}] \Rightarrow (\mathit{adore} \cdot x \cdot \mathit{john})$$

Thus, intuitively, if an interpreter knows that a pantechnicon is a vehicle, she will be able to bind the presupposition triggered by the definite *the vehicle* in (5).

Now suppose the interpreter does *not* know that a pantechnicon is a vehicle. That is, Γ does not contain a function mapping pantechnicons to vehicles. Then, still under the assumption that Γ itself does not introduce any vehicles, the interpreter will not be able to prove the existence of a vehicle. Intuitively this means that the interpreter is faced with an expression containing an unsatisfied presupposition. In that case, she might come to the conclusion that her context is not rich enough and that something (namely a vehicle) is missing from it. She can then try to *accommodate* the existence of a vehicle by replacing the gaps Y and P with fresh variables, say y' and p', and extending the context Γ with $y' : \mathit{entity}, p' : \mathit{vehicle} \cdot y'$. Of course, it has to be checked whether this move is adequate, whether the accommodation is consistent, etc.

4 BRIDGING

Let us take stock. We claim that context, and more in particular, world knowledge plays a role in presupposition projection. However, there are very few, if any, theories of presupposition which account for the interaction between presuppositions and context/world knowledge. We have argued that the deductive perspective of CTT offers an attractive framework to model this interaction. Our starting point is a reformulation of Van der Sandt's presupposition resolution algorithm tailored to CTT. In this section we want to illustrate the formal interaction between world knowledge and presupposition resolution, by focusing on the bridging phenomenon.[5]

So, what *is* bridging precisely? Clark [1975] describes it in terms of an interpreter who is looking for an antecedent, but cannot find one 'directly in memory'. "*When this happens, he is forced to construct an antecedent, by a series of inferences, from something he already knows. (...) The listener must therefore bridge the gap from what he knows to the intended antecedent.*" [Clark, 1975, p. 413]. We want to make these general ideas more precise. In particular, we want to spell out the notion of inference that is involved. Clark himself contends that the bridging-inferences are similar in nature to what Grice [1975] has called 'implicatures'. From the current perspective, there are two kinds of inferences relevant for bridging. The most straightforward one would simply be inference in CTT. We take it that a CTT context Γ represents the information an agent has 'directly in memory'. Inferred information corresponds with objects that can be constructed from objects in Γ using the deduction rules of CTT. However, there is also a second kind of inference present in the approach to presuppositions sketched above: accommodation (which bears a close resemblance to *abduction* in the framework of [Hobbs *et al.*, 1993; Krause, 1995]). We claim that both kinds of inference play a role in bridging. Let us discuss each in somewhat more detail.

'Inference' as Deduction in CTT

From this perspective, bridging amounts to using world knowledge to fill gaps. Consider first example (10.a) with its CTT representation given in (10.b).

(10) a. If John buys a car, he checks the motor first.

b. $[x : \textit{entity}, y : \textit{car} \cdot x, z : \textit{buy} \cdot x \cdot \textit{john}] \Rightarrow$
$$(\textit{check} \cdot Y \cdot \textit{john})_{[Y:\textit{entity}, P:\textit{motor} \cdot Y]}$$

Before we can add this expression to some context Γ, we have to resolve the presuppositional expression. To do so, we first search for a substitution S such that the following can be proven:

(11) $\Gamma, x : \textit{entity}, y : \textit{car} \cdot x, z : \textit{buy} \cdot x \cdot \textit{john} \vdash (Y : \textit{entity}, P : \textit{motor} \cdot Y)[S]$

Let us assume that Γ (a model of the agent's 'direct memory') does not contain a sufficiently salient motor. Then the interpreter will try to 'bridge the gap from what he knows to the intended antecedent'. When does he succeed in this, i.e. when can *the motor* be understood as a bridging anaphor

[5]Elsewhere (in [Krahmer and Piwek, to appear]) we have shown that the CTT approach also yields interesting results for the interaction between presupposition projection in conditionals and world knowledge.

licensed by the introduction of a car? The answer is simple: if the interpreter knows that a car has a motor. Modelling this knowledge could go as follows. Γ contains two functions: one function which maps each car to an entity, $f : ([a : \textit{entity}, b : \textit{car} \cdot a] \Rightarrow \textit{entity})$, and one function which states that this entity is the car's motor $g : ([a : \textit{entity}, b : \textit{car} \cdot a] \Rightarrow (\textit{motor} \cdot (f \cdot a \cdot b))$. Using these functions, we find a substitution S in (11), mapping Y to $f \cdot x \cdot y$ and P to $g \cdot x \cdot y$. We can look at the resulting proof objects as the 'bridge' that has been constructed by the interpreter; it makes the link with the introduction of a car explicit (by using x and y) and indicates which inference steps the user had to make to establish the connection with the motor (by using the functions f and g). Thus, we can fill the gaps, assuming that the proofs satisfy certain conditions. Of course, they have to satisfy the usual Van der Sandt conditions (such as consistency). Additionally, the bridge itself has to be 'plausible'. Below we will return to the issue of constraints on building bridges.

'Inference' as Accommodation

Let us now consider a somewhat more complex example (after Clark [1975, p. 416]).

(12) John walked into the room. The chandelier sparkled brightly.

Let us assume that the first sentence of (12) has already been processed, which means that the context Γ contains the following introductions: $x : \textit{entity}, y : \textit{room} \cdot x, z : \textit{walk\_in} \cdot x \cdot \textit{john}$. At this stage, we want to deal with the CTT representation of the second sentence, given below.

(13) $q : \textit{sparkle} \cdot Y_{[Y : \textit{entity}, P : \textit{chandelier} \cdot Y]}$

We want to resolve the presupposition triggered by *the chandelier* in the context Γ (assuming that Γ does not introduce any (salient) chandeliers). When would an interpreter be able to link *the chandelier* to the room John entered? Of course, it would be easy if she had some piece of knowledge to the effect that every room has a chandelier (if her Γ contained functions which for each room produce a chandelier). However, such knowledge is hardly realistic; many rooms do not have a chandelier.

In a more lifelike scenario, the following might happen. The interpreter tries to prove the existence of a chandelier, but fails to do so. However, the interpreter knows that a chandelier is a kind of lamp and the existence of a lamp *can* be proven using the room just mentioned and the background knowledge that rooms have lamps. Formally, and analogous to the 'motor' example, one function which produces an entity for each room; $f : ([a : \textit{entity}, b : \textit{room} \cdot a] \Rightarrow \textit{entity})$, and one which states that this entity is a lamp; $g : ([a : \textit{entity}, b : \textit{room} \cdot a] \Rightarrow (\textit{lamp} \cdot (f \cdot a \cdot b)))$. Since the speaker has uttered (12) the interpreter will *assume* that (one of) the lamp(s) in the room is a chandelier (compare [Clark, 1975, p. 416]). In terms of the CTT approach, this could go as follows: first, the interpreter *infers* (deduces) that the room which John entered contains an entity which is a lamp (applying the aforementioned piece of knowledge; the functions f and g), and then binds *part* of the presupposition by filling the Y gap with $f \cdot x \cdot y$ (the inferred lamp). The remaining part of the presupposition (that the lamp is in fact a chandelier) is now *accommodated* in the Van der Sandtian way by filling the gap with a fresh variable.

Bridging as a Determinate Process

Clark [1975] claims that bridging is a *determinate* process. In theory, however, background knowledge will license a number of bridges. In CTT terms, there will often not be *one* way to fill a presuppositional gap, but there will be many. Clark discusses the following example:

(14) Alex went to a party last night. He is going to get drunk again tonight.

Here *again* triggers the presupposition that Alex was drunk before. According to Clark, we assume that "*every time Alex goes to a party, he gets drunk*". In our opinion, this assumption is too strong, we feel that one would merely assume that Alex was drunk at the party he visited last night (compare the 'chandelier' case). But that is not the point here. Clark [1975, pp. 419–420] goes on to notice that there are theoretically conceivable alternatives for his assumption which interpreters, however, would never construct: "(...) *we could have assumed instead that every time he* [Alex, P&K] *goes to party he meets women, and all women speak in high voices, and high voices always remind him of his mother, and thinking about his mother always makes him angry, and whenever he gets angry, he gets drunk*" We would like to stress that the problem of determinacy is not restricted to bridging. Consider the following example from Lewis [1979, p. 348]:

(15) The pig is grunting, but the pig with the floppy ears is not grunting.

Apparently, this sentence can only be uttered when there are (at least) two pigs in 'direct memory'. Nevertheless, each of the definite descriptions can be understood as referring to a determinate pig. Lewis argues that *salience* is the relevant notion here: he argues that *the pig* is the most salient pig, while *the pig with the floppy ears* is the most salient pig with floppy ears. In other words: the interpreter has to find the most salient antecedent for the respective descriptions in order to guarantee determinedness. However, in the case of bridging, salience is a necessary, but certainly not a sufficient condition to guarantee determinedness. We propose to use two (groups of) conditions to come to a *determinate* bridge, related to the *effort* an interpreter needs to construct a bridge and the ***plausibility*** of the constructed bridges.

The Effort Condition

To begin with the former: as noted above, Clark [1975, p. 420] claims that interpreters do not draw inferences *ad infinitum*, and to model this he proposes a general *stopping rule*, which says essentially that the interpreter builds the shortest possible bridge that is consistent with the context. We take this constraint to subsume two conditions. The first of these conditions boils down to the following rule: *use your 'informational resources' as frugal as possible.* In CTT terms: if a gap can be filled with more than one proof object, fill it with the one with the lowest complexity.[6] As a (rather informal) illustration, consider:

(16) A married couple is strolling through the park. The man looks in love.

Let us assume that the interpreter has evaluated the first sentence and added its representation to Γ. Thus, Γ contains an introduction for a married couple, say x. Now say that the interpreter has (at least) the following background knowledge: 'every married couple contains a man' (call this function f_1), 'every married couple contains a woman' (f_2) and 'everyone has a father' (g), assuming that fathers are male. If the interpreter now looks for an antecedent for the presupposition triggered by the man, she will find $f_1 \cdot x$ (the male half of the married couple), but also $g \cdot (f_1 \cdot x)$ (the father of the male half of the married couple), not to mention $g \cdot (f_2 \cdot x)$ (the father of the female half of the married couple). However, the proof-object $f_1 \cdot x$ is clearly less complex (makes thriftier use of the 'informational resources') than the last two proof-objects, and therefore the interpreter uses this object to fill the presuppositional gap, and not the other two.

The second condition subsumed by Clark's stopping rule can be put as follows: *make as few assumptions as possible* (accommodate as little as possible). Example (14) provides an illustration of this second condition, which

[6] The complexity of a proof object is defined as the number of unbound variables in the proof object. A variable which occurs as part of a proof object and which is not bound by a λ operator corresponds to an object in Γ. [λ abstraction basically introduces a hypothetical object into a proof, e.g. we can proof $p \to q$ from q by starting with a proof a for q and then introducing a hypothetical proof x for p: $(\lambda x : p.a) : (p \to q)$. Thus we obtain a function, which when it is applied to some proof y of p yields a proof for q, namely $(\lambda x : p.a) \cdot y = a$.] Thus, our complexity measure takes into account the amount of information needed from Γ to construct the proof-object and also how many times an object has been used. Assume for instance that $\Gamma = p : prop, q : prop, a : p, f : p \to q$ (in words, p and q are propositions, and a is a proof of p and f is a proof of $p \to q$, respectively). We denote the complexity of object o by $C(o)$. For the proof a of p we have: $C(a) = 1$. For q we have $C(f \cdot a) = 2$. Note that for tautologies there are proofs whose complexity is equal to zero (we need no premisses to proof them).

on the current approach follows from the general view of accommodation as a repair strategy (modelled by a preference for resolution/binding over accommodation). Factors like recency are also related to minimising effort; it takes more effort to build a bridge to an antecedent occurring ten sentences ago then to one occurring in the previous sentence.

The Plausibility Condition

Suppose that we can fill the gap associated with a bridging anaphor with two objects which are indistinguishable under the effort-conditions: the utterance is ambiguous between two different assertions. However, if one of the assertions is less 'plausible' than the other, this helps us to select a determinate reading. Consider the following mini-dialogue:

(17) *John*: Why did Tom drive Mike's car and not his own?
Bill : Because the motor had broken down.

John's question presupposes that Tom drove Mike's car and not his own. The description *the motor* can either be licensed by Mike's car or by Tom's car. If John uses the background knowledge that one cannot drive a car with a broken motor, he will be able to derive an inconsistency from Bill's answer (combined with the presupposition of his question) when he takes Mike's car as an 'antecedent' for *the motor*, but not when *the motor* is part of Tom's car. Based on consistency requirements, only the interpretation of Bill's utterance which answers John's question is selected. We feel that relevance (e.g. answerhood) often is a *side-effect* of consistency, which we take to be a minimal condition of plausibility. Another illustration of the plausibility condition is the following:

(18) Mary traded her old car in for a new one. The motor was broken.

In this example, the motor can be licensed by both Mary's old car and by her new car. Nevertheless, one has a very strong tendency to interpret *the motor* as referring to Mary's old car. To check these intuitions, we conducted a small enquiry via email among Dutch subjects with (a Dutch version) of (18) as well as some other examples.[7] After reading the example, the subjects were presented with the following query: *Of which car was the motor broken?* ◯ *The old one, or* ◯ *The new one.* Subjects were asked to provide the first answer that came to their mind. The results were unequivocal: 2 subjects choose the new car, while 48 interpreted *the motor* as referring to Mary's old car. How can we account for this? One possibility

[7]The questionnaire (consisting of four multiple-choice questions) was returned by 50 native speakers of Dutch working at IPO. It included participants from a number of different backgrounds: (computational) linguists, perception scientists (vision, auditory), physicists, computer scientists, psychologists, ergonomists, secretaries and management.

is this: the 48 subjects have the 'knowledge' that a new car has a working motor. If an interpreter has such background knowledge in her Γ, than she will be able to derive an inconsistency when the description the motor is linked to Mary's new car. Hence this resolution is not plausible, and thus rejected in favour of the other resolution.[8]

In the two examples discussed so far, two possible readings remained after applying the effort condition, one of which could later be ruled out due to plausibility (consistency) with the side-effect that the selected reading is 'relevant' (i.e. provides an explanation for the event described in the first sentence). However, it can also happen that exactly one reading remains under the effort conditions, although this reading is rather implausible, e.g. it is inconsistent with the world-knowledge of the interpreter. Consider example (19), from [Asher and Lascarides, 1996, p. 16].

(19) a. I met two interesting people last night who voted for Clinton.
 b. The woman abstained from voting in the election.

The only available antecedent for the woman in (19.b) is one of the two people mentioned in (19.a). As Asher and Lascarides point out, the only available binding reading "(...) ***results in an inconsistency that makes the discourse sound strange***". Nevertheless, this (inconsistent) reading is the preferred (only) one. They use this example to indicate a difference with the abductive framework proposed in Hobbs *et al.* [1993]. In that framework, presuppositional and asserted material are treated on a par. As a consequence Hobbs *et al.* predict a reading in which an antecedent for the woman is accommodated. Here our approach makes the same prediction as Asher and Lascarides' approach. The plausibility condition selects the most plausible reading from the readings which passed the effort condition (those readings requiring least effort). Obviously, if only one reading survived the effort condition, it is by definition the most plausible reading.

The picture that emerges is one where interpretation involves two stages: first a stage where some readings (if any) are selected on the basis of effort and then a second stage in which the interpreter selects the most plausible reading. Accommodation is only an option if the effort condition yields no binding reading whatsoever. Note that this approach gives a particular meaning to the idea that an interpreter tries to make sense of what has been said. The process of making sense is constrained by simple effort conditions. In other words, although for a particular utterance there may

[8] Notice that it can happen that both the effort and the plausibility conditions put together fail in selecting one most preferred proof- object. In that case an unresolvable ambiguity results; no determinate bridge can be constructed. The following provides an illustration of this: ?? ***If John buys a car and a motorbike, he'll check the engine first.*** There are two potential antecedents for the presupposition triggered by the engine which are indistinguishable under both the effort and the plausibility condition. As a result, the sentence is odd (marked by the double question mark).

be theoretically possible readings that 'make sense' (e.g. are 'explanatory' with respect to the preceding discourse), these readings may simply not be available, because there are other readings which, although they make less sense, present themselves more readily to the interpreter. Let us illustrate this with another example. Consider:

(20) John moved from Brixton to St. John's Wood.
The rent was less expensive.

Matsui [1995] found that people interpret the 'rent as anaphoric to the rent in St. John's Wood. What is more, Matsui's experiment showed that this preference even overrides the default knowledge present in the subjects of the experiment that the rents are generally more expensive in St. Johns Wood than in Brixton. Asher and Lascarides [1996, p. 10] argue that: "(...) *intuitively, one prefers explanations of intentional changes (in this case, moving house) to simple background information that sets the scene for the change*". If *the rent* is anaphoric to the rent in St. John's Wood, we get an explanation for John's removal. But are they right? Consider:

(21) John moved from Brixton to St. John's Wood.
The rent was more expensive.

If we apply the Asher and Lascarides analysis to this example ('explanation preferred'), then the prediction is that an interpreter should prefer a reading where *the rent* refers to the rents in Brixton, since *the rents in Brixton being more expensive than those in St. John's Wood* would provide a good reason for John's removal. In the aforementioned experiment we tested Asher and Lascarides' prediction with example (22):[9]

(22) John moved from Horst to Maasbree. The rent was more expensive.

After reading this example, subjects where asked the following question: *Where is the rent more expensive? ◯Horst, or ◯Maasbree.* Again subjects were asked to provide the first answer that came to their mind. The results for this particular question were as follows: Horst: 8, Maasbree: 42. These results ($\chi^2 = 23.12, p < 0.001$) are the opposite of Asher and Lascarides' prediction: *the rent* is interpreted as the rent of John's new domicile in Maasbree/St. John's wood.

Let us now sketch an alternative explanation in terms of our framework. The basic idea is that the effort condition suggests the most salient antecedent as the most likely candidate. In this case, salience is influenced by the temporal interpretation of the two sentences. In particular, the notion of a reference event, as proposed in [Hinrichs, 1986] may play a central role.

[9]Horst and Maasbree are two small Dutch towns which are not very well known.

In line with the rules provided by Hinrichs, this reference event will be located immediately after the moving event. In other words, for the reference event it holds that John is living in Maasbree and thus he is paying the rent in Maasbree. According to Hinrichs, if a sentence expresses a state, then the reference event of the previous sentence should be temporally included in this state. Thus, the state expressed by 'The rent was more expensive' should include the referent event where John lives in Maasbree. This is what makes the reading where *the rent* refers to the rent in Maasbree the most preferred reading.[10] As regards to plausibility, neither the reading where *the rent* refers to the rent in Horst nor the one where it refers to the rent in Maasbree is implausible (leads to an inconsistency), thus the reading preferred by the effort condition is retained. Note that the same story can be told for Asher and Lascarides' example. Sometimes, plausibility can override the ordering suggested by effort:

(23) John moved from Horst to Maasbree. The rent was too expensive.

In this case, the reading where John moves to a place where the rent is *too* expensive for him has little plausibility; background knowledge like 'one cannot pay things which are too expensive' will rule out this reading.

5 RELATED WORK

In this section we compare our proposals with some related work from the literature. Various authors have studied presuppositions from a proof-theoretic perspective. In [Ranta, 1994], CTT is extended with rules for definite descriptions. Ahn [1994] and Beun and Kievit [1995] use CTT for dealing with the resolution of definite expressions. The latter focus on selecting the right referent (which can come not only from the linguistic context, but also from the physical context) using concepts such as prominence and agreement. Krause (1995) presents a type-theoretical approach to presupposition. His theory not only allows binding of presuppositions, but also has the possibility to globally accommodate them using an abductive inferencing mechanism. One important difference with our approach is that we simply take the *entire* theory of Van der Sandt and rephrase it in terms of CTT. Apart from that, we also want to show that this reformulation paves the way for a formalized influence of background knowledge on presupposition projection (in particular: bridging). In general, we believe that the CTT approach leads to better results than adding a proof system to DRT, as done in e.g. [Saurer, 1993]. The main advantage of CTT is that it is a standard proof system developed in mathematics with well-understood

[10] A somewhat comparable approach is advocated in [Poesio, 1994]. Poesio shows how shifts in the focus of attention influence the interpretation of definite descriptions.

meta-theoretical properties. And, as we have shown, the presence of explicit proof objects in CTT has some additional advantages for the treatment of bridging.

We are aware of three formal approaches to bridging: the abductive approach [Hobbs, 1987; Hobbs *et al.*, 1993], the lexical approach ([Bos *et al.*, 1995] and the rhetorical approach [Asher and Lascarides, 1996].

Abduction: Chandeliers Revisited

We have analysed example (12) in terms of CTT deduction and accommodation, where the latter is similar to the notion of implicature argued for by Clark. An analysis of (12) in terms of implicature has also been presented in [1987]. Though our approach is similar to his in spirit, it differs in the details. Hobbs suggest the following basic scheme for implicature: *IF P is mutually known, (P&R) → Q is mutually known, and the discourse requires Q, THEN assume R as mutually known and CONCLUDE Q.* In case of the chandelier example, Q is instantiated with 'there is a chandelier', P with 'there is a lamp' and R with 'in the form of a branching fixture'. Uttering the first sentence makes it mutually known that there is a lamp. Hobbs now explains the use of the definite *the chandelier* in the second sentence as follows: the interpreter 'accommodates' that this lamp has the form of a branching fixture, and thus can derive the presence of a chandelier.

Let us compare this to the present approach. We also assume that the first sentence introduces a lamp. And similarly, the definite in the second sentence requires the presence of a chandelier. Furthermore, we assume that it is part of the world knowledge of the interlocutors that chandeliers are lamps. The idea is now that the interpreter is licensed to assume that the lamp in question is a chandelier in virtue of (*i*) the fact that the linguistic context contains a salient lamp, (*ii*) the fact that chandeliers are lamps and (*iii*) the fact that a chandelier is presupposed. One important difference is that we do not require a decomposition of chandelier into a lamp in the form of a branching fixture. A further difference with the approach of Hobbs *et al.* is that on our approach it is the independently motivated presupposition resolution algorithm which drives the bridging process.

Is Bridging a Lexical Phenomenon?

Bos [1995] treat bridging as a *lexical* phenomenon. They combine a version of Van der Sandt's presupposition resolution algorithm with a *generative lexicon* [Pustejovky, 1995]. In this way each potential antecedent for a presupposition is associated with a *qualia-structure* indicating which 'concepts' can be associated with the antecedent. As they put it, a qualia-structure can be seen as a set of lexical entailments. Our main objection to this approach is that not all implied antecedents are *lexical* entailments. The examples in

(24) illustrate the importance of non-lexical background knowledge.

(24) a. Yesterday Chomsky analysed a sentence on the blackboard, but I couldn't see the tree.
b. Yesterday somebody parked a car in front of my door, and the dog howled awfully.

For most people trees have as much to do with sentences as dogs have with cars. Yet, both these examples can have a bridging reading, given a suitable context. The a. example requires some basic knowledge concerning formal grammars which most readers of this paper presumably will have. For them, (24.a) is a perfectly normal thing to say under its bridging reading (because all of them have a mental function mapping sentences to trees). Likewise, (24.b) can be understood in a bridging manner given the 'right' background knowledge. Suppose, it is well-known between speaker and interpreter that the former lives opposite a home for stray animals somewhere in the countryside, and all cars which stop in front of this home for lost animals either drop a dog or pick one up. In this context, the interpreter will have no trouble constructing the required bridge (since she has a mental function which produces a dog for each car stopping in front of the speaker's door).

Bridging as a Byproduct of Computing Rhetorical Structure

In the previous section we already discussed the work of Asher and Lascarides [1996], who claim that bridging is determined by rhetorical structure. We have illustrated that the predictions of Asher and Lascarides in connection with examples (20)–(22) are not in accord with the results we obtained in a small observational study. This is not to say that we disagree with the basic tenet from Asher and Lascarides that discourse structure, and in particular, rhetorical relations between sentences play a factor in bridging. For example, we believe that their observation concerning examples like *John was going to commit suicide. He got a rope.* [Charniak, 1983] is correct; the fact that the second sentence is somehow constructed as 'background' to the first may be useful to infer that the rope is going to be used for the suicide (essentially this is due to Grice's [1975] maxim of relevance). We do not agree, however, with the claim that rhetorical structure is the driving force behind bridging (that it is a "*byproduct of computing rhetorical structure*", [Asher and Lascarides, 1996, p. 1]. In our opinion, rhetorical structure is certainly an influence on bridging, but we believe that world knowledge, as well as factors like effort and plausibility play a more substantial role, as we have tried to argue in connection with examples (20)–(22).

6 CONCLUSION

We have discussed a deductive variant of Van der Sandt's presuppositions as anaphora theory. In this new perspective presuppositions are treated as gaps, which have to be filled using contextual information. This information can come from the linguistic context, that is the preceding sentences (as in [Van der Sandt, 1992]). But presuppositional gaps can also be filled using the non-linguistic context, i.e. world knowledge. As an illustration of the formal interaction between world knowledge and presupposition, we have applied our deductive approach to Clark's bridging cases. We distinguished two cases: if the 'bridge' between presupposition and would-be antecedent is fully derivable using context (including world knowledge), the presupposition associated with the anaphor can be bound. This means that binding plays a more substantial role than in Van der Sandt's original theory, as presuppositions can be bound to both inferred and non-inferred antecedents. If the 'bridge' between anaphor and antecedent is not fully derivable, the 'missing link' will be accommodated. So, accommodation is still a repair strategy, as in Van der Sandt's original approach, but now there is generally less to repair. In most cases, accommodation will amount to 'assuming' that an object of which the existence has been proven satisfies a more specific description (in the case of (12), that the lamp whose existence has been proven on the basis of *a room* is in fact a chandelier).

Our approach to bridging resembles the abductive approach advocated by Hobbs *et al.* [1993], but there are also a number of important differences. We take it that the presuppositionhood is the driving force behind bridging. As a consequence we make a strict separation between presupposed and asserted material, thereby avoiding the problems Hobbs and co-workers have with examples like (19). Additionally, we are not committed to lexical decomposition, such as chandeliers are lamps with a branching fixture. The knowledge that chandeliers are lamps is sufficient. We have tried to argue that bridging is not a purely lexical process (as opposed to [Bos *et al.*, 1995]), and that bridging is not a byproduct of rhetorical structure determination (as opposed to [Asher and Lascarides, 1996]). This is not to say that lexical matters or rhetorical relations have no relevance for bridging, we feel that they are two of the *many* factors which play a role in bridging.

It is well known that bridging is not an unrestricted process. Therefore we have tried to formulate two general constraints/filters on the bridging process partly inspired by the informal observations from [Clark, 1975]: *effort* (the 'mental capacity' the interpreter needs to construct a bridge) and *plausibility* (the relative admissibility of the constructed bridges). We have modelled the effort constraint in CTT terms as follows: if a gap can be filled by more than one proof object, order them with respect to proof complexity (the one with the lowest complexity first). Moreover, use the 'informational resources' as frugal as possible (accommodate as little as possible). This latter condition is 'hard-coded', as it were, in the Van der Sandtian resolution algorithm. We take it that factors like recency and salience are also relevant here. The plausibility condition is modelled as a simple consistency condition, with relevance as a side-effect. We have seen that these two simple conditions help in arriving at a determinate bridge. The precise formulation of these two conditions and the interplay between them will be the subject of further research. Additionally, we are interested in further empirical validation of our predictions along the lines of the observational study described in this paper.

ACKNOWLEDGEMENTS

We would like to thank René Ahn, Nicholas Asher, Kees van Deemter and Jerry Hobbs for discussion and comments.

Paul Piwek
ITRI - University of Brighton, Brighton, UK.

Emiel Krahmer
Eindhoven University of Technology, Eindhoven, The Netherlands.

REFERENCES

[Ahn, 1994] R. Ahn. Communicating Contexts: A Pragmatic Approach to Information Exchange. In: P. Dybjer, B. Nordström and J. Smith (eds.), *Types for Proofs and Programs: selected papers*. Springer Verlag. Berlin, 1994.

[Ahn and Kolb, 1990] R. Ahn and H-P. Kolb. Discourse Representation meets Constructive Mathematics. In: L. Kálmán and L. Pólos (eds.), *Papers from the second Symposium on Logic and Language*. Budapest, 1990.

[Asher and Lascarides, 1996] N. Asher and A. Lascarides. Bridging. In: R. Van der Sandt, R. Blutner and M. Bierwisch (eds.), *From Underspecification to Interpretation*, IBM Deutschland, Heidelberg, 1996. Also in *Journal of Semantics*, 15(1): 83–113, 1998.

[Barendregt, 1992] H. P. Barendregt. Lambda Calculi with Types. In: S. Abramsky, D. Gabbay and T. Maibaum (eds.), *Handbook of Logic in Computer Science*, Oxford University Press, 1992.

[Beaver, 1997] D. Beaver. Presupposition. In: J. Van Benthem and A. Ter Meulen (eds.), *The Handbook of Logic and Language*, pp. 939–1008, Elsevier, 1997.

[Beun and Kievit, 1995] R. J. Beun and L. Kievit. *Resolving definite descriptions in dialogue*. manuscript/DenK-report. Tilburg/Eindhoven, 1995.

[Bos *et al.*, 1995] J. Bos, P. Buitelaar and M. Mineur. Bridging as Coercive Accommodation. In: E. Klein *et al.* (eds.), *Working Notes of the Edinburgh Conference on Computational Logic and Natural Language Processing*. HCRC, 1995.

[Bunt, 1995] H. Bunt. Dialogue Control Functions and Interaction Design. In: R. J. Beun, M. Baker and M. Reiner (eds.), *Dialogue in Instruction*, pp. 197–214, Springer Verlag, Berlin, 1995.

[Charniak, 1983] E. Charniak. Passing Markers: A Theory of Contextual Influence in Language Comprehension. In: *Cognitive Science* **7**, 171– 190, 1983.

[Clark, 1975] H. Clark. Bridging. In: R. Schank and B. Nash-Webber (eds.), *Theoretical Issues in Natural Language Processing*, MIT Press, Cambridge, MA, 1975. Quotes from reprint in P. N. Johnson-Laird and P. C. Wason (eds.), *Thinking*. Cambridge University Press, 411–420.

[Curry and Feys, 1958] H. B. Curry and R. Feys. *Combinatory Logic*. Vol. 1, North-Holland, Amsterdam, 1958.

[Grice, 1975] H. Grice. Logic and Conversation. In: P. Cole and J. L. Morgan (eds.), *Syntax and Semantics 3: Speech Acts*, Academic Press, New York, 1975.

[Hinrichs, 1986] E. Hinrichs. Temporal Anaphora in Discourses of English, *Linguistics and Philosophy* **9**: 63–82, 1986.

[Hobbs, 1987] J. Hobbs. *Implicature and Definite Reference*. Stanford: Report CSLI-87-99, 1987.

[Hobbs *et al.*, 1993] J. Hobbs, M. Stickel, D. Appelt and P. Martin. Interpretation as Abduction. In: *Artificial Intelligence*, **63**: 69–142, 1993.

[Kamp and Reyle, 1993] H. Kamp and U. Reyle. *From Discourse to Logic*. Kluwer Academic Publishers, Dordrecht, 1993.

[Krahmer, 1998] E. Krahmer. *Presupposition and Anaphora*. CSLI Publications, Stanford, 1998.

[Krahmer and Piwek, to appear] E. Krahmer and P. Piwek. Presupposition Projection as Proof Construction. In: H. Bunt and R. Muskens (eds) *Computing Meanings: Current Issues in Computational Semantics*. Kluwer Academic Publishers, Dordrecht, to appear.

[Krause, 1995] P. Krause. Presupposition and Abduction in Type Theory. In: E. Klein *et al.* (eds.), *Working Notes of the Edinburgh Conference on Computational Logic and Natural Language Processing*, HCRC, Edinburgh, 1995.

[Lewis, 1979] D. Lewis. Scorekeeping in a Language Game. In: *Journal of Philosophical Logic* **8**:339–359, 1979.

[Martin-Löf, 1984] P. Martin-Löf. *Intuitionistic Type Theory*. Bibliopolis, Naples, 1984.

[Matsui, 1995] T. Matsui. *Bridging and Relevance*. Ph.D. thesis, University of London, 1995.

[Piwek, 1997] P. Piwek. The Construction of Answers. In: A. Benz and G. Jäger (eds.), *Proceedings of MunDial: the München Workshop on the Formal Semantics and Pragmatics of Dialogue*. CIS-Bericht, University of Munich, 1997.

[Piwek, 1998] P. Piwek. *Logic, Information and Conversation*. Ph.D. thesis, Eindhoven University of Technology, 1998.

[Pustejovky, 1995] J. Pustejovky. *The Generative Lexicon*. MIT Press, Cambridge, MA, 1995.

[Poesio, 1994] M. Poesio. Definite Descriptions, Focus Shift and a Theory of Discourse Interpretation. In: R. van der Sandt and P. Bosch (eds.), *Proceedings of the Conference on Focus in Natural Language Processing*, IBM, Heidelberg, 1994.

[Ranta, 1994] A. Ranta. *Type-theoretical Grammar*. Clarendon Press, Oxford, 1994.

[Saurer, 1993] W. Saurer. A natural deduction system for DRT. *Journal of Philosophical Logic* **22**: 249–302, 1993.

[Stalnaker, 1974] R. Stalnaker. Pragmatic Presuppositions. In: M. Munitz and P. Unger (eds.), *Semantics and Philosophy*, pp. 197–213, New York University Press, New York, 1974.

[Van der Sandt, 1992] R. Van der Sandt. Presupposition Projection as Anaphora Resolution. *Journal of Semantics* **9**:333-377, 1992.

VAGAN Y. TERZIYAN AND SEPPO PUURONEN

REASONING WITH MULTILEVEL CONTEXTS IN SEMANTIC METANETWORKS

1 INTRODUCTION

It is generally accepted that knowledge has a contextual component. Acquisition, representation, and exploitation of knowledge in context would have a major contribution in knowledge representation, knowledge acquisition, and explanation, as Brezillon and Abu-Hakima supposed in [Brezillon and Abu-Hakima, 1995]. Among the advantages of the use of contexts in knowledge representation and reasoning Akman and Surav [Akman and Surav, 1996] mentioned the following: economy of representation, more competent reasoning, allowance for inconsistent knowledge bases, resolving of lexical ambiguity and flexible entailment. Brezillon and Cases noticed however in [Brezillon and Cases, 1995] that knowledge-based systems do not use correctly their knowledge. Knowledge being acquired from human experts does not usually include its context.

Contextual component of knowledge is closely connected with eliciting expertise from one or more experts in order to construct a single knowledge base (or, for example as in [Brezillon, 1994], for cooperative building of explanations). Could the overlapping knowledge, obtained from multiple sources, be described in such a way that it becomes context or even process independent? In [1995], Taylor *et al.* gives the negative answer. Certainly there have been inference engines produced that were subsequently applied to related domains, but in general the sets of rules have been different. If more than one expert are available, one must either select the opinion of the best expert or pool the experts' judgements. It is assumed that when experts' judgements are pooled, collectively they offer sufficient cues leading to the building of a comprehensive theory.

Very important questions have been raised in [Brezillon and Abu-Hakima, 1995]. Does context simplify or complicate the construction of a knowledge base? Is context an object of the domain? Can we move from one context to the next one [McCarthy, 1993]? What are the possible formalisms that seem to allow the explicit representation of a context?

McCarthy [McCarthy, 1993] illustrates how a reasoning system can utilize contexts to incorporate information from a general common-sense knowledge base into other specialized knowledge bases. The basic relation he uses is the $ist(c, p)$, which asserts that the proposition p is true in the context c. He introduces contexts as abstract mathematical entities with properties and relations. One remark is that it would be useful to have a formal theory of the natural use of the logic for representation. Later in [McCarthy, 1995] it was noticed that statements about contexts are themselves in contexts. Contexts may be treated as mathematical structures of

P. Bonzon, M. Cavalcanti and R. Nossum (eds.), Formal Aspects of Context, 107–126.

different properties and they also may have relations with other contexts. According to McCarthy there is no general context in which all the stated axioms always hold.

Different uses of contexts were analyzed by Guha in [Guha, 1991]. In his approach all axioms and statements are not universally true, they are only true in contexts which he calls microtheories that make different assumptions about the world. He has developed some very general lifting rules with which different microtheories can be integrated. He also uses contexts for integrating multiple databases and handling mutual inconsistencies between databases.

Buvac *et al.* investigated the simple logical (in [Buvac *et al.*, 1995; Buvac *et al.*, 1993]) and semantic (in [Buvac *et al.*, 1994]) properties of contexts. This formal theory of context and the use of quantificational logic enables the presentation of relations between contexts, operations on contexts, and state lifting rules of facts in different contexts. The quantificational logic of context [Buvac, 1996], for example, enables to state that the formula ψ is true in all contexts which satisfy some property $p(x)$ as follows: $(\forall v)p(v) \rightarrow ist(v, \psi)$. Each context is considered to have a set of propositional atoms as its own vocabulary. It was shown that the acceptance of outermost context simplifies the metamathematics of the contexts.

Attardi and Simi [1995] offer a viewpoint representation of context related formalisms. Viewpoints denote sets of statements which represent the assumptions of a theory. The basic relation in their formalization: $in(\text{'}A\text{'}, vp)$ where vp is a viewpoint expression, can be interpreted as 'statement A is entailed by the assumption denoted by vp'. Operations between viewpoints are carried out with metalevel rules. The relation $Holds(A, s) = in(A, vp(s))$, which defines the set of facts $vp(s)$ holding in a situation s, is used to handle situations as sets of basic facts.

Edmonds [1997] describe a simple extension of semantic nets. The nodes are labeled and the directed arcs can lead to other arcs as well as nodes. In this model, contexts are not differentiated as special objects, but rather that some nodes to a greater or lesser extent have roles as encoders of contextual information. This formulation is shown to be expressive enough to capture several aspects of contexts including reasoning and generalization in contexts. It is not claimed that this is a model of any type of context found in human activity.

After making their comparison of approaches towards formalizing context, Akman and Surav [1996] concluded that the idea of formalizing context seems to have caught on and produce good theoretical outcomes and the area of innovative applications remains relatively unexplored.

As one can imagine looking context related references that the context itself has so many different meanings relatively to the goals, application areas, formal methods, and so on. To understand and apply the research results in the context area one has to be involved to the context in which these results have been obtained. The goal of this paper is to view the world of contexts, the properties, relationships, and rules of contexts and their context for interpretation. The context of contexts (metacontext) can be considered recursively as several levels of contexts, the topmost being a *universal context*. This universal context is supposed to include knowledge which always remains the same in every context. The representation of such multilevel structure requires special multilevel semantic formalism which we call a *semantic metanetwork.*

The idea of a semantic metanetwork is generally acknowledged by Puuronen and Terziyan in [1992] and further developed in [Bondarenko *et al.*, 1993; Terziyan, 1993]. In [Puuronen and Terziyan, 1992] the objects of the higher level representation were treated as 'birth' and 'death' rules for relations between objects of the lower level. Thus the semantic metanetwork together with the description of the basic domain contains the rules of change in time. Sometimes it is also necessary to change the rules. This is done by the metarules of the next level of a semantic metanetwork. The more complex the dynamics of the described domain is, the more metalevels the semantic metanetwork contains.

This paper discusses another use of a semantic metanetwork. The zero level of representation is a semantic network that includes knowledge about basic domain objects and their relations in various contexts. The first level of representation uses a semantic network to present knowledge about contexts and their relationships. Relationships among contexts are considered in metacontexts. The second level of representation defines relationships between metacontexts, and similarly for each next level. The topmost level includes knowledge which is considered to be 'truth' in every context.

What are the main uses of semantic metanetworks? The need of metaknowledge is broadly acknowledged in knowledge engineering and its main use is to help inference (metareasoning has been dealt for example in [Russel and Wefald, 1991]). Shastri presents in his article [Shastri, 1989] two important kinds of inference: inheritance and recognition. Woods [Woods, 1991] discusses very deeply the classification of a concept with respect to a given taxonomic structure. Shapiro [Shapiro, 1991] discussed also inference as a reduction inference and a path-based inference. A context can be considered as a mechanism for reasoning, too. One advantage of making context explicit in a representation is the capability to inference within and across contexts, and thus explicitly make the changes in reasoning across contexts [McCarthy, 1993].

Our goal is to present methods of reasoning using knowledge in its context which is formally represented by semantic metanetworks. The problems are: how to derive knowledge about any relation which is interpreted in the highest contextual level; how to derive the initial relation when the context in which the relation has been interpreted and the result of interpretation are known; how to derive knowledge about an unknown context by analyzing its effect to the initial knowledge; and how to transform knowledge from one context to another.

2 A SEMANTIC METANETWORK

In this section we define a semantic metanetwork formally and give an example.

A semantic metanetwork is a formal representation of knowledge with contexts.

The knowledge of a basic domain is represented by a semantic network with nodes (domain objects) and arcs (relations between domain objects). There are two types of relations: a relation between two domain objects (nodes) and a relation of a domain object with itself (an arc from a node to itself). The second type of the relation presents a property of an object.

We also consider a context as a domain object. Sometimes it is necessary to describe knowledge about contexts, too. Knowledge about properties and relations of contexts compose a new semantic network which is considered to describe the next level of knowledge representation. Levels of such representation have relationships with each other.

Formally a metanetwork is a quadruple $\langle A, L, S, D\rangle$, where A is a set of objects which consists of the subset of the basic domain objects $A_i^0, i = 1, \ldots, n_1$, and several subsets of the contexts $A_i^{(d)}, i = 1, \ldots, n_d$, and $d = 1, \ldots, klev$ identifies the level of the metanetwork where the context appears; L is a set of unique names of relations $L_k^{(d)}, k = 1, \ldots, m_d$ and $d = 0, \ldots, klev$ identifies the level of the metanetwork where the relation appears; S is the set of relations $S_r^{(d)} = P(A_i^{(d)}, L_k^{(d)}, A_j^{(d)}), r = 1, \ldots, l_d$ composed in each level so that: $S^{(d)} = \wedge_r S_r^{(d)}$, and $P(A_i^{(d)}, L_k^{(d)}, A_j^{(d)})$ is true when there is the relation $L_k^{(d)} \in L$ between the objects $A_i^{(d)}$ and $A_j^{(d)}, (A_i^{(d)}, A_j^{(d)} \in A)$ at the level d; and D is the set of context predicates $D_r^{(d)} = ist(A_i^{(d+1)}, S_r^{(d)})$ connecting the contexts of the level $d+1$ to the relations of the level d and $ist(A_i^{(d+1)}, S_r^{(d)})$ is true if the relation $S_r^{(d)}$ holds in the context $A_i^{(d+1)}$.

Let us consider an example of a semantic metanetwork presented in Figure 1. The metanetwork in Figure 1 can be described by the following expressions:

$$A = \{A^0, A', A''\}, A^0 = \{A_1, A_2, A_3\},$$
$$A' = \{A'_1, A'_2, A'_3, A'_4\}, A'' = \{A''_1, A''_2, A''_3\}$$

which are the set of the objects and its three subsets at the different levels;

$$L = \{L^0, L', L''\}, L^0 = \{L_1, L_2, L_3, L_4\},$$
$$L' = \{L'_1, L'_2, L'_3\}, L'' = \{L''_1, L''_2\}$$

which are the names of the relations and its three subsets at the different levels;

$$S = \{S^0, S', S''\}, S^0 = \{S_1, S_2, S_3, S_4\},$$
$$S' = \{S'_1, S'_2, S'_3\}, S'' = \{S''_1, S''_2\}$$

which are the set of relations and its three subsets at the different levels;

$$S_1 = P(A_1, L_1, A_2), S_2 = P(A_2, L_2, A_3), S_3 = P(A_1, L_3, A_3),$$
$$S_4 = P(A_3, L_4, A_3); S'_1 = P(A'_1, L'_1, A'_2), S'_2 = P(A'_2, L'_2, A'_3),$$
$$S'_3 = P(A'_2, L'_3, A'_4); S''_1 = P(A''_1, L''_1, A''_2), S''_2 = P(A''_2, L''_2, A''_3)$$

which are the relations of all the three levels in a predicate form;

$$D = \{D^0, D'\}, D^0 = \{D_1, D_2, D_3, D_4, D_5\}, D' = \{D'_1, D'_2, D'_3\}$$

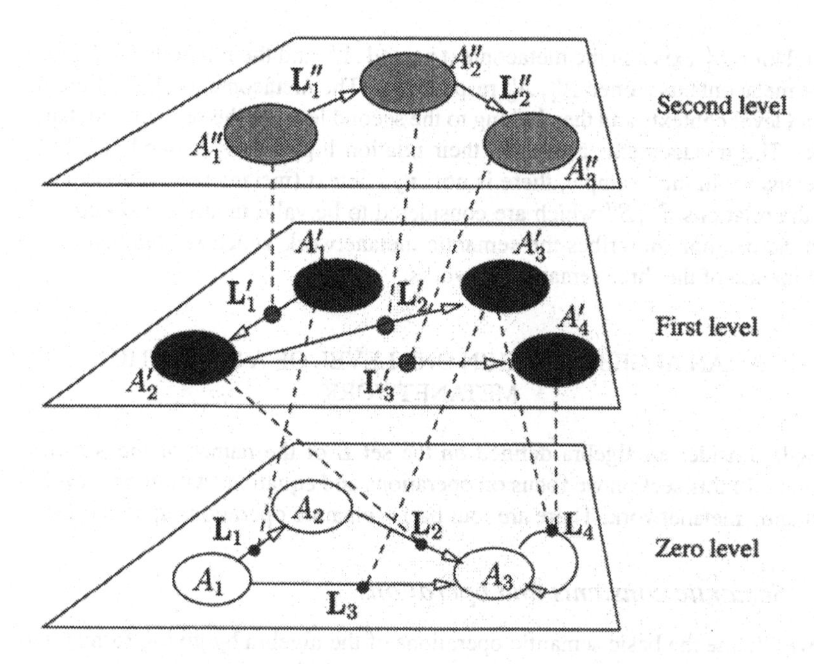

Figure 1. An example of a semantic metanetwork

which are the set of the context predicates and its two subsets at the different levels;

$$D_1 = ist(A'_1, S_1), D_2 = ist(A'_2, S_2), D_3 = ist(A'_3, S_3),$$
$$D_4 = ist(A'_3, S_4), D_5 = ist(A'_4, S_4)$$

which are the context predicates describing the context relationships between the zero and the first level;

$$D'_1 = ist(A''_1, S'_1), D'_2 = ist(A''_2, S'_2), D'_3 = ist(A''_3, S'_3)$$

which are the context predicates describing the context relationships between the first and the second level.

In Figure 1 and in the formal description of the example, the objects A_1, A_2, A_3 are the basic domain objects of the basic level of the semantic metanetwork. The basic domain relations are: S_1 between objects A_1 and A_2 named as L_1; S_2 between objects A_2 and A_3 named as L_2; S_3 between objects A_1 and A_3 named as L_3; S_4 the object A_3 with itself named as L_3 (thus L_3 is the name of the property of the object A_3). The relation S_1 exists in the context named A'_1, and the relations S_2–S_3 exist in the contexts named A'_2–A'_3 respectively. The relation S_4 exists if the two contexts A'_3, A'_4 hold. The contexts A'_1–A'_4 are the first level contexts and they belong to the first level of the semantic metanetwork. In the example, the first level contexts also have their relationships S'_1–S'_3 named as L'_1–L'_3 respectively.

The relation S_1' exists in the metacontext named A_1'', and the relations S_2', S_3' exist in the metacontexts named A_2'', A_3'' respectively. The metacontexts A_1''–A_3'' are the second level contexts and they belong to the second level of the semantic metanetwork. The metacontexts also have their relationships S_1'', S_2'' named as L_1'', L_2'' respectively. In the example, there is not any context (metametacontext) to interpret the relations S_1'', S_2'' which are considered to be valid in any known context. Thus the example describes the semantic metanetwork which is obtained as the combination of the three semantic networks.

3 AN ALGEBRA WITHIN ONE LEVEL OF A SEMANTIC METANETWORK

We will consider an algebra defined on the set L of the names of the semantic relations. In this section we focus on operations and equations within one level of a semantic metanetwork. There are four basic semantic operations upon the L-set.

3.1 Semantic constants and operations

We will define the basic semantic operations of the algebra by giving formal definition and properties and also by giving a graphical interpretation. Further in formulas we will use '$\Leftrightarrow$' to mark the logical equivalence between two predicates, '$\Rightarrow$' to mark the logical consequence between two predicates, '$\equiv$' between the names of relations or objects to mark the equality of the semantic meanings of two names, '$\neq$' between the names of the relations to mark the inequality of the semantic meanings of two names. Proofs for some of properties are also presented. We use, with formal definitions of operations, the following notation:

$$(P(A_i, L_k, A_j) \Leftrightarrow P(A_i, L_m, A_j)) \Leftrightarrow L_k \equiv L_m.$$

Semantic constants

It is supposed that two objects represented in the semantic network by two different circles with two different names A_i and A_j are different objects. Even if there is no knowledge about a relationship between the two objects, then at least it is known that they are different objects. We consider knowledge about the difference as a semantic constant *DIF* which denotes the relation between any pair of different objects:

$$\forall A_i, A_{j(j \neq i)}(P(A_i, DIF, A_j) \Leftrightarrow \mathit{true}).$$

On the other hand the above formula can be interpreted so that there is no difference between any object with itself. This statement defines another semantic constant of equivalence *SAME*:

$$\forall A_i(P(A_i, SAME, A_i) \Leftrightarrow \mathit{true}).$$

The last formula also means that if there is not any knowledge about the properties of an object, then at least one property always holds—to be same to itself.

We define *NIL* relation as a relation which never holds between any two objects or among the properties of any object by the following way:

$$\forall A_i, A_{j(j \neq i)}(\neg P(A_i, DIF, A_j) \Leftrightarrow P(A_i, NIL, A_j) \Leftrightarrow false),$$
$$\forall A_i(\neg P(A_i, SAME, A_i) \Leftrightarrow P(A_i, NIL, A_i) \Leftrightarrow false).$$

When the semantic network is being built the following assumption is used. If there exist two different pieces of knowledge and there is not explicitly known that they belong to the description of the same object then it is assumed that they describe different objects. Sometimes it happens that semantics (properties and relationships) of these two objects are always the same. When this happens then the two objects should be connected by the binary equivalent of the *SAME* relation using the following definition:

$$(P(A_s, L_k, A_i) \Leftrightarrow P(A_s, L_k, A_j), \forall A_s \in A, \forall L_k \in L) \Leftrightarrow P(A_i, SAME^b, A_j),$$

where $SAME^b$ is the binary equivalent of the *SAME* relation. This relation does not mean the negation of the *DIF* relation, because it means only that the two different objects have the same semantics.

Semantic inversion

Formally the semantic inversion can be defined by the following equation:

$$P(A_i, L_k, A_j) \Leftrightarrow P(A_j, \tilde{L}_k, A_i),$$

where $\tilde{L}_k$ is the new inverse relation. It will be named with the unique symbol L_m and it is added to the set L. For example, if $L_k = \langle to\_punish \rangle$, then $L_m = \langle to\_be\_punished \rangle$ and $L_m \equiv \tilde{L}_k$. Similarly, let $L_n = \langle to\_be\_on\_the\_left\_side\_of \rangle$, then $L_m = \langle to\_be\_on\_the\_right\_side\_of \rangle$ and $L_m \equiv \tilde{L}_n$.

If the relation in the definition means a property of an object then its inversion is equal to the original relation: $P(A_i, L_k, A_i) \Leftrightarrow P(A_i, \tilde{L}_k, A_i)$ which means the unary relation: $L_k \equiv \tilde{L}_k$.

The obvious property for the semantic inversion operation is *double inversion*: $\tilde{\tilde{L}}_k \equiv L_k$.

Semantic negation

The semantic negation operation means changing the name of a relation if the value of an appropriate predicate is false. This is defined in a following way:

$$\neg P(A_i, L_k, A_j) \Leftrightarrow P(A_i, \bar{L}_k, A_j),$$

where $\bar{L}_k$ is the new relation. It is named with the unique symbol L_m and it is added to the set L. If, for example, $P(\langle Mary \rangle, \langle to\_love \rangle, \langle Tom \rangle) = false$, then it means the same as: $P(\langle Mary\rangle, \langle not\_to\_love\rangle, \langle Tom\rangle) = true$. Thus, if $L_k = \langle to\_love\rangle$ and $L_m = \langle not\_to\_love\rangle$, then $L_m \equiv \bar{L}_k$.

Properties

- double negation: $\bar{\bar{L}}_k \equiv L_k$;
- negation of inversion: $\bar{\tilde{L}}_k \equiv \tilde{\bar{L}}_k$;

Proof.

$$(P(A_i, \bar{\tilde{L}}_k, A_j) \Leftrightarrow \neg P(A_i, \tilde{L}_k, A_j) \Leftrightarrow \neg P(A_j, L_k, A_i) \Leftrightarrow$$
$$P(A_j, \bar{L}_k, A_i) \Leftrightarrow P(A_i, \tilde{\bar{L}}_k, A_j)) \Leftrightarrow \bar{\tilde{L}}_k \equiv \tilde{\bar{L}}_k.$$

■

Semantic multiplication

The semantic multiplication operation defines the name of an unknown relation between a pair of objects A_i and A_j if there exists the third object A_s which is connected with them both. The formal definition is as follows:

$$P(A_i, L_k, A_s) \wedge P(A_s, L_n, A_j) \Leftrightarrow P(A_i, L_k * L_n, A_j),$$

where $L_k * L_n$ is the new relation. It is named with the unique symbol L_m and it is added to the set L. If, for example, it is true that $P(\langle Mary\rangle, \langle to\_be\_married\_with\rangle, \langle Tom\rangle)$ and $P(\langle Tom\rangle, \langle to\_have\_mother\rangle, \langle Diana\rangle)$, then it is also true that: $P(\langle Mary\rangle, \langle to\_have\_mother\text{-}in\text{-}law\rangle, \langle Diana\rangle)$. Thus, if $L_k = \langle to\_be\_married\_with\rangle$, $L_n = \langle to\_have\_mother\rangle$, and $L_m = \langle to\_have\_mother\text{-}in\text{-}law\rangle$, then $L_m \equiv L_k * L_n$.

Properties

- non-commutativity
  $\neg(L_k * L_n \equiv L_n * L_k), \forall L_k, L_n (L_k \neq L_n \neq DIF \neq SAME^b)$.
- $L_k * (L_n * L_m) \equiv (L_k * L_n) * L_m$;

Proof.

$$(P(A_i, L_k * (L_n * L_m), A_j) \Leftrightarrow$$
$$P(A_i, L_k, A_s) \wedge P(A_s, L_n * L_m, A_j) \Leftrightarrow$$
$$P(A_i, L_k, A_s) \wedge$$
$$P(A_s, L_n, A_t) \wedge P(A_t, L_m, A_j) \Leftrightarrow$$
$$P(A_i, L_k * L_n, A_t) \wedge P(A_t, L_m, A_j) \Leftrightarrow$$
$$P(A_i, (L_k * L_n) * L_m, A_j)) \Leftrightarrow$$
$$L_k * (L_n * L_m) \equiv (L_k * L_n) * L_m.$$

■

- inversion over multiplication $\sim(L_k * L_n) \equiv \tilde{L}_n * \tilde{L}_k$;

Proof.

$$P(A_i, \sim(L_k * L_n), A_j) \Leftrightarrow$$
$$P(A_j, L_k * L_n, A_i) \Leftrightarrow P(A_j, L_k, A_s) \wedge P(A_s, L_n, A_i) \Leftrightarrow$$
$$P(A_s, \tilde{L}_k, A_j) \wedge P(A_i, \tilde{L}_n, A_s) \Leftrightarrow P(A_i, \tilde{L}_n, A_s) \wedge P(A_s, \tilde{L}_k, A_j) \Leftrightarrow$$
$$P(A_i, \tilde{L}_n * \tilde{L}_k, A_j)) \Leftrightarrow \sim(L_k * L_n) \equiv \tilde{L}_n * \tilde{L}_k.$$

■

Semantic addition

The semantic addition operation defines the name of the relation between a pair of objects A_i and A_j as a combination of two relations between these two objects. Formally:

$$P(A_i, L_k, A_j) \wedge P(A_i, L_n, A_j) \Leftrightarrow P(A_i, L_k + L_n, A_j),$$

where $L_k + L_n$ is the new relation. It is named with the unique symbol L_m and it is added to the set L. If, for example, it is true that $P(\langle Mary \rangle, \langle to\_give\_birth\_to \rangle, \langle Tom \rangle)$ and $P(\langle Mary \rangle, \langle to\_take\_care\_of \rangle, \langle Tom \rangle)$, then: $P(\langle Mary \rangle, \langle to\_give\_birth\_to\_and\_take\_care\_of \rangle, \langle Tom \rangle)$ is also true. Thus, if $L_k = \langle to\_give\_birth\_to \rangle$, $L_n = \langle to\_take\_care\_of \rangle$, and $L_m = L_k + L_n \langle to\_give\_birth\_to\_and\_take\_care\_of \rangle$.

It is also possible to sum properties. If, for example, it is true that $P(\langle Tom \rangle, \langle to\_be\_clever \rangle, \langle Tom \rangle)$ and $P(\langle Tom \rangle, \langle to\_be\_rich \rangle, \langle Tom \rangle)$, then it is also true that: $P(\langle Tom \rangle, \langle to\_be\_clever\_and\_rich \rangle, \langle Tom \rangle)$.

Properties

- commutativity $L_k + L_n \equiv L_n + L_k$;
- transitivity $L_k + (L_n + L_m) \equiv (L_k + L_n) + L_m$;
- reflexivity $L_k + L_k \equiv L_k$;
- inversion over sum $\sim(L_k + L_n) \equiv \tilde{L}_k + \tilde{L}_n$;
- distributivity (left) $L_k * (L_m + L_n) \equiv L_k * L_m + L_k * L_n$;

Proof.

$$(P(A_i, L_k * (L_m + L_n), A_j) \Leftrightarrow$$
$$P(A_i, L_k, A_s) \wedge P(A_s, L_m + L_n, A_j) \Leftrightarrow$$
$$P(A_i, L_k, A_s) \wedge P(A_s, L_m, A_j) P(A_s, L_n, A_j) \Leftrightarrow$$
$$P(A_i, L_k, A_s) \wedge P(A_i, L_k, A_s) P(A_s, L_m, A_j) \wedge P(A_s, L_n, A_j)) \Leftrightarrow$$
$$P(A_i, L_k * L_m, A_j) \wedge P(A_i, L_k * L_n, A_j) \Leftrightarrow$$
$$P(A_i, L_k * L_m + L_k * L_n, A_j) \Leftrightarrow$$
$$L_k * (L_m + L_n) \equiv L_k * L_m + L_k * L_n.$$

■

- distributivity (right) $(L_k + L_m) * L_n \equiv L_k * L_n + L_m * L_n$.

Semantic closeness

We define the L-set $L : \{L_1, L_2, \ldots, L_n\}$ as semantically closed set of the names of the relations if the result of any semantic operation with any operands from the L-set also belongs to the L-set. Formally we define semantic closeness as follows:

$$\begin{gathered}
\tilde{L}_k \in L, \forall L_k \in L; \bar{L}_k \in L, \forall L_k \in L; \\
L_k * L_m \in L, \forall L_k, L_m \in L; \\
L_k + L_m \in L, \forall L_k, L_m \in L; \\
\tilde{L} : \{\tilde{L}_1, \tilde{L}_2, \ldots, \tilde{L}_n\} \equiv L; \\
\bar{L} : \{\bar{L}_1, \bar{L}_2, \ldots, \bar{L}_m\} \equiv L; \\
(L_i * L) : \{L_i * L_1, L_i * L_2, \ldots, L_i * L_n\} \equiv L, \forall L_i \in L; \\
(L * L_i) : \{L_1 * L_i, L_2 * L_i, \ldots, L_n * L_i\} \equiv L, \forall L_i \in L; \\
(L_i + L) : \{L_i + L_1, L_i + L_2, \ldots, L_i + L_n\} \equiv L, \forall L_i \in L.
\end{gathered}$$

In the following text we will consider the L-set as semantically closed one.

Operations with semantic constants

In the semantically closed L-set there should be relations *DIF*, *NIL* and *SAME* as they were defined in Section 3.1.1.

This means that at least one of the possible relations should take place between a pair of objects and that each object has at least one property. The negation operation defines the main relationship between the two main constants as follows:

$$\overline{DIF} \equiv \overline{SAME} \equiv NIL.$$

Properties of DIF

- neutrality to semantic sum $\quad DIF + L_k \equiv L_k$.

Proof.

$$(P(A_i, DIF + L_k, A_j) \Leftrightarrow P\underbrace{(A_i, DIF, A_j)}_{true} \wedge P(A_i, L_k, A_j) \Leftrightarrow$$

$$\Leftrightarrow P(A_i, L_k, A_j)) \Leftrightarrow DIF + L_k \equiv L_k.$$

■

- elimination in semantic multiplication $\quad DIF * L_k \equiv L_k * DIF \equiv DIF$;
- inversion of *DIF* $\quad {}^{\sim}(DIF) \equiv DIF$;

Properties of SAME

- inversion of *SAME* $\sim(SAME) \equiv SAME$;
- neutrality to semantic sum $SAME + L_k \equiv L_k$;

Properties of $SAME^b$

- inversion of $SAME^b$ $\sim(SAME^b) \equiv SAME^b$;
- neutrality to semantic sum $SAME^b + L_k \equiv L_k$;
- neutrality to semantic multiplication $L_k * SAME^b \equiv SAME^b * L_k \equiv L_k$.
- annihilation

  a) $L_k + \tilde{L}_k \equiv \begin{cases} SAME^b, & \text{for relations;} \\ L_k, & \text{for properties.} \end{cases}$

  b) $L_k * \tilde{L}_k \equiv \tilde{L}_k * L_k \equiv SAME^b$.
- elimination (left) $L_k + L_k * L_m \equiv L_k * L_m$.

Proof.

$$L_k + L_k * L_m \equiv L_k * SAME^b + L_k * L_m \equiv L_k * (SAME^b + L_m) \equiv L_k * L_m.$$

■

3.2 *Semantic equations*

An equation of the algebra (a semantic equation) is the equality of two expressions which include only the names of the semantic relations as operands and only semantic operations with these operands. In a semantic equation there should be at least one unknown operand. We use the following general expression to denote an equation: $F_1(L^1) = F_2(L^2)$, where F_1, F_2 are functions, which use operations of the algebra and they have known operands respectively subsets L^1, L^2 of the semantically closed L-set. Here and later we use '$=$' in semantic equations to make the difference between semantic equations and semantic identities where we use '$\equiv$'. We will consider equations with one unknown operand L_x, which is defined on the whole L-set.

The solution of such equation is the relation L_w, which, being substituted in an equation instead of the operand L_x will transform it to an identity. We will use notation $(equation)^{L_x=L_w}$ as a statement that relation L_w is the solution of the *equation*.

To solve semantic equations the following rules are used:

$$(F_1(L^1) = F_2(L^2))^{L_x=L_w} \Leftrightarrow (\tilde{F}_1(L^1) = \tilde{F}_2(L^2))^{L_x=L_w};$$
$$(F_1(L^1) = F_2(L^2))^{L_x=L_w} \Leftrightarrow (\bar{F}_1(L^1) = \bar{F}_2(L^2))^{L_x=L_w};$$
$$(F_1(L^1) = F_2(L^2))^{L_x=L_w} \Leftrightarrow (F_1(L^1) + L_k = F_2(L^2) + L_k)^{L_x=L_w}, \forall L_k \in L;$$
$$(F_1(L^1) = F_2(L^2))^{L_x=L_w} \Leftrightarrow$$
$$(F_1(L^1) * L_k = F_2(L^2) * L_k)^{L_x=L_w}, \forall (L_k \neq DIF \in L);$$
$$(F_1(L^1) = F_2(L^2))^{L_x=L_w} \Leftrightarrow$$
$$(L_k * F_1(L^1) = L_k * F_2(L^2))^{L_x=L_w}, \forall (L_k \neq DIF \in L).$$

Using the above rules and properties of semantic operations, we present methods of the decision of the following five types of equations of the algebra.

$$\tilde{L}_x = L_i \Rightarrow \tilde{\tilde{L}}_x = \tilde{L}_i \Rightarrow L_x = \tilde{L}_i; \quad \bar{L}_x = L_i \Rightarrow \bar{\bar{L}}_x = \bar{L}_i \Rightarrow L_x = \bar{L}_i;$$
$$L_x + L_i = L_j \Rightarrow L_x + L_i + \tilde{L}_i = L_j + \tilde{L}_i \Rightarrow$$
$$L_x + SAME^b = L_j + \tilde{L}_i \Rightarrow L_x = L_j + \tilde{L}_i;$$
$$L_x * L_i = L_j \Rightarrow L_x * L_i * \tilde{L}_i = L_j * \tilde{L}_i \Rightarrow$$
$$L_x * SAME^b = L_j * \tilde{L}_i \Rightarrow L_x = L_j * \tilde{L}_i;$$
$$L_i * L_x = L_j \Rightarrow \tilde{L}_i * L_i * L_x = \tilde{L}_i * L_j \Rightarrow$$
$$SAME^b * L_x = \tilde{L}_i * L_j \Rightarrow L_x = \tilde{L}_i * L_j.$$

4 AN ALGEBRA WITH CONTEXTS IN A SEMANTIC METANETWORK

In this chapter we will further develop the algebra, described in the previous chapter, by adding a new semantic operation which allows to reason with the multilevel structure of contexts in a semantic metanetwork.

4.1 Operation of semantic interpretation

Semantic interpretation operation makes it possible to take into account the properties of the appropriate context when one defines knowledge of a relationship between a pair of objects. If we have knowledge L_k about the name of a relation between the objects A_i and A_j, and knowledge L'_n about the name of the property of the context of this relation (as shown in Figure 2a), then we can obtain a new knowledge about this relation which is supposed to be an interpretation of the first knowledge in the context of the second one. By the same way it is possible to interpret knowledge about a property of an object in a certain context as it is shown in Figure 2b.
Formally:

$$P(A_i, L_k, A_j) \wedge P(A'_l, L'_n, A'_l) \wedge ist(A'_l, P(A_i, L_k, A_j)) \Leftrightarrow P(A_i, L_k^{L'_n}, A_j),$$

where $L_k^{L'_n}$ denotes the interpretation of the knowledge L_k using the knowledge L'_n about the context A'_l. This is named and included to the set L. We use a

a)

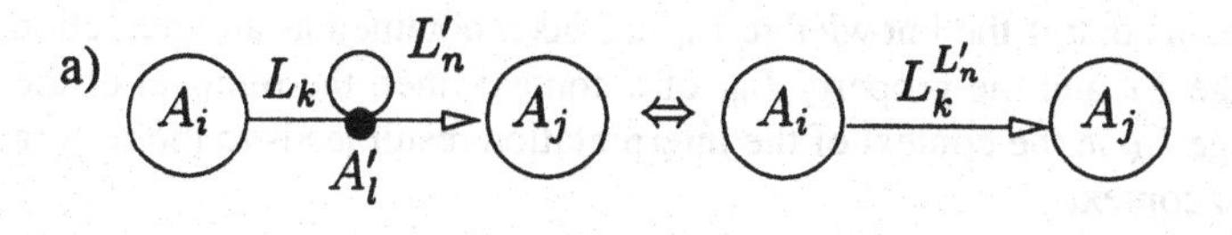

b)

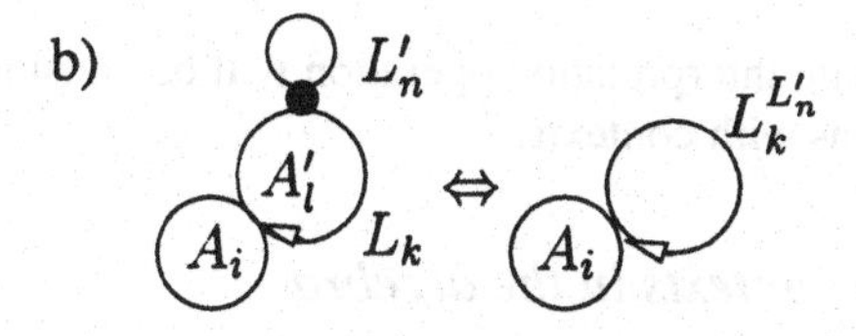

Figure 2. Semantic interpretation

property of a context in the description of our basic operation which connects different levels of a semantic metanetwork assuming that different contexts with the same properties have the same effect to the interpretation of knowledge.

We expand the definition of the relations *SAME* and $SAME^b$, defined in the previous chapter, adding the following properties:

1. $L_k^{SAME} \equiv L_k$,

which means an abstract case when the context of a relation contains only the universal property *SAME*. In such a case the context cannot change an interpretation of the relation;

2. $SAME^{L_k} \equiv L_k$,

which means also an abstract case when the knowledge being interpreted contains only the universal property *SAME*. In such a case the only knowledge obtained as a result of interpretation is the knowledge about context;

3. $(SAME^b)^{L_k} \equiv L_k^b$,

which means that interpretation of the equivalence relation between two objects inherits the properties of the context producing its binary equivalent L_k^b;

4. $(L_k^{L_m} \equiv L_n) \Leftrightarrow (L_n^{\bar{L}_m} \equiv L_k)$—axiom of extracting knowledge,

which means that if some knowledge L_n has been obtained as an interaction of the knowledge L_k and the property L_m of a context, then the removal of that property from the context of the interpretation result leads to the restoration of the initial knowledge;

Consequence $L_k^{\bar{L}_k} \equiv SAME^b$;

5. $(L_k^{L_m} \equiv L_n) \Leftrightarrow (\bar{L}_k^{L_n} \equiv L_m)$—axiom of extracting context,

which means that if the knowledge L_n has been obtained as an interaction of the knowledge L_k and the property L_m of a context, then the removal of the initial knowledge L_k in the context of the interpretation result leads to the restoration of the initial context.

Consequence $\bar{L}_k^{L_k} \equiv SAME^b$.

Other properties of the semantic interpretation operation will be considered in the next section as transformations with contexts.

4.2 *Transformations with contexts in the algebra*

In this paragraph, we describe the rules of transformation of the algebra expressions which include contexts.

Inversion in the context. The inverse result of the interpretation of a relation in a context is equal to the result of the interpretation of the inverse relation in the same context. Formally:

$$\sim(L_k^{L_m}) \equiv \tilde{L}_k^{L_m}.$$

Negation in the context. The negative result of the interpretation of a relation in a context is equal to the result of the interpretation of the negative relation in the same context, and it is also equal to the result of the interpretation of this relation in a negative context. Formally:

$$\overline{L_k^{L_m}} \equiv \bar{L}_k^{L_m} \equiv L_k^{\bar{L}_m}.$$

Addition in the context. The result of the interpretation of the sum of two not-conflicting relations in a context is equal to the semantic sum of these two relations interpreted separately in the same context. Formally:

$$(L_k + L_m)^{L_n} \equiv L_k^{L_n} + L_m^{L_n}, \text{ when } L_k \neq \bar{L}_m.$$

Multiplication in the context. The result of the interpretation of the semantic multiplication of two not-conflicting relations in a context is equal to the semantic multiplication of these two relations interpreted separately in the same context. Formally:

$$(L_k * L_m)^{L_n} \equiv L_k^{L_n} * L_m^{L_n}, \text{ when } L_k \neq \bar{L}_m.$$

Interpretation in the sum of contexts. The result of the interpretation of a relation in a semantic sum of two not-conflicting contexts is equal to the semantic sum of this relation interpreted separately in these two contexts. Formally:

$$L_k^{L_m+L_n} \equiv L_k^{L_m} + L_k^{L_n}, \text{ when } L_m \neq \bar{L}_n.$$

Addition of interpreted relations. The result of a semantic sum of two not-conflicting relations interpreted separately in two not-conflicting contexts is equal to the semantic sum of these relations interpreted in the semantic sum of these contexts:

$$L_k^{L_m} + L_r^{L_n} \Rightarrow (L_k + L_r)^{L_m + L_n}, \text{ when } L_k \neq \bar{L}_r, L_m \neq \bar{L}_n.$$

Multiplication of interpreted relations. The result of a semantic multiplication of two not-conflicting relations interpreted separately in two not-conflicting contexts is equal to the semantic multiplication of these relations interpreted in the semantic sum of these contexts:

$$L_k^{L_m} * L_r^{L_n} \Rightarrow (L_k * L_r)^{L_m + L_n}, \text{ when } L_k \neq \bar{L}_r, L_m \neq \bar{L}_n.$$

Multilevel interpretation. The result of a relation interpretation in several levels of contexts does not depend on the order of interpretation:

$$(L_k^{L_m})^{L_n} \equiv L_k^{(L_m^{L_n})}, \text{ when } L_m \neq \bar{L}_k, L_m \neq \bar{L}_n.$$

4.3 Equations with contexts in the algebra

To solve equations with the operation of semantic interpretation, the following main rules are used:

$$(F_1(L^1) = F_2(L^2))^{L_x = L_w} \Leftrightarrow (F_1(L^1)^{L_k} = F_2(L^2)^{L_k})^{L_x = L_w}, \forall L_k \in L;$$
$$(F_1(L^1) = F_2(L^2))^{L_x = L_w} \Leftrightarrow (L_k^{F_1(L^1)} = L_k^{F_2(L^2)})^{L_x = L_w}, \forall L_k \in L.$$

Using the above properties and rules, we present solution methods of the following two types of equations of the algebra.

$$L_x^{L_i} = L_j \Leftrightarrow (L_x^{L_i})^{\bar{L}_i} = L_j^{\bar{L}_i} \Leftrightarrow L_x^{(L_i^{\bar{L}_i})} = L_j^{\bar{L}_i} \Leftrightarrow$$
$$L_x^{SAME} = L_j^{\bar{L}_i} \Leftrightarrow L_x = L_j^{\bar{L}_i};$$
$$L_i^{L_x} = L_j \Leftrightarrow \bar{L}_i^{(L_i^{L_x})} = \bar{L}_i^{L_j} \Leftrightarrow (\bar{L}_i^{L_i})^{L_x} = \bar{L}_i^{L_j} \Leftrightarrow$$
$$SAME^{L_x} = \bar{L}_i^{L_j} \Leftrightarrow L_x = \bar{L}_i^{L_j}.$$

Using the above more complex equations can be solved. For example:

$$\begin{aligned}
&\tilde{L}_1 * L_2^{(\tilde{L}_3 * \tilde{L}_x * L_4 + L_5)^{\bar{L}_6}} * L_7 + L_8 = L_9 \Leftrightarrow \\
&\tilde{L}_1 * L_2^{(\tilde{L}_3 * \tilde{L}_x * L_4 + L_5)^{\bar{L}_6}} * L_7 = L_9 + \tilde{L}_8 \Leftrightarrow \\
&L_2^{(\tilde{L}_3 * \tilde{L}_x * L_4 + L_5)^{\bar{L}_6}} * L_7 = \tilde{\tilde{L}}_1 * (L_9 + \tilde{L}_8) \Leftrightarrow \\
&L_2^{(\tilde{L}_3 * \tilde{L}_x * L_4 + L_5)^{\bar{L}_6}} = L_1 * (L_9 + \tilde{L}_8) * \tilde{L}_7 \Leftrightarrow \\
&L_2^{(\tilde{L}_3 * \tilde{L}_x * L_4 + L_5)} = (L_1 * (L_9 + \tilde{L}_8) * \tilde{L}_7)^{\bar{\bar{L}}_6} \Leftrightarrow \\
&\tilde{L}_3 * \tilde{L}_x * L_4 + L_5 = \bar{L}_2^{(L_1 * (L_9 + \tilde{L}_8) * \tilde{L}_7)^{\bar{\bar{L}}_6}} \Leftrightarrow \\
&\tilde{L}_3 * \tilde{L}_x * L_4 = \bar{L}_2^{(L_1 * (L_9 + \tilde{L}_8) * \tilde{L}_7)^{\bar{\bar{L}}_6}} + \tilde{L}_5 \Leftrightarrow \\
&\tilde{L}_x * L_4 = \tilde{\tilde{L}}_3 * (\bar{L}_2^{(L_1 * (L_9 + \tilde{L}_8) * \tilde{L}_7)^{\bar{\bar{L}}_6}} + \tilde{L}_5) \Leftrightarrow \\
&\tilde{L}_x = L_3 * (\bar{L}_2^{(L_1 * (L_9 + \tilde{L}_8) * \tilde{L}_7)^{\bar{\bar{L}}_6}} + \tilde{L}_5) * \tilde{L}_4 \Leftrightarrow \\
&L_x = {\sim}(L_3 * (\bar{L}_2^{(L_1 * (L_9 + \tilde{L}_8) * \tilde{L}_7)^{\bar{\bar{L}}_6}} + \tilde{L}_5) * \tilde{L}_4).
\end{aligned}$$

5 REASONING WITH A SEMANTIC METANETWORK

Using semantic metanetworks and equations of the algebra, one can solve the four reasoning problems with contexts described in the following subsections.

Deriving an interpreted knowledge (decontextualization). Deriving an interpreted knowledge means deriving the formula for knowledge interpreted using all the known levels of its context. This problem is solved using operations of the algebra.

We define the knowledge of a relation $L_{A_i - A_j}$ between any pairs of objects (A_i, A_j) from the same level of a semantic metanetwork as a semantic sum over all possible paths between these objects (A_i, A_j) at this level of the metanetwork.

We define the knowledge L_{A_i} of an object A_i of a semantic metanetwork in one level as a semantic sum over all knowledge of the relations that connect this object with all objects of the same level, including the object itself:

$$L_{A_i} = \sum_j L_{A_i - A_j}.$$

The interpreted knowledge of any relation, considering all contexts and metacontexts, is derived by the following schema:

$$\begin{aligned}
&\langle \text{interpreted knowledge} \rangle = \\
&\quad \langle \text{knowledge} \rangle^{\langle \text{knowl.aboutcontext} \rangle^{\cdots \langle \text{knowl.aboutmetacontextofnthlevel} \rangle}}
\end{aligned}$$

As an example let us derive the interpreted knowledge of the relation between A_1 and A_3 in Figure 1. We will start from the top level of the metanetwork and define knowledge about the metacontexts A''_1, A''_2, A''_3:

$$\begin{gathered} L_{A''_1} = L_{A''_1-A''_2} + L_{A''_1-A''_3} + L_{A''_1-A''_1} = \\ = L''_1 + L''_1 * L''_2 + SAME = L''_1 * (SAME + L''_2) = L''_1 * L''_2; \\ L_{A''_2} = \tilde{L}''_1 + L''_2; \text{ and } L_{A''_3} = \tilde{L}''_2 * \tilde{L}''_1. \end{gathered}$$

Now we can continue at the first level of the metanetwork and derive the interpreted knowledge of the first level relations:

$$\begin{gathered} L_{A'_1-A'_2} = (L'_1)^{L_{A''_1}} = (L'_1)^{L''_1 * L''_2}; \\ L_{A'_1-A'_3} = (L'_1)^{L_{A''_1}} * (L'_2)^{L_{A''_2}} = (L'_1)^{L''_1 * L''_2} * (L'_2)^{\tilde{L}''_1 + L''_2} = \\ (L'_1 * L'_2)^{L''_1 * L''_2 + \tilde{L}''_1}; \\ L_{A'_1-A'_4} = (L'_1)^{L_{A''_1}} * (L'_3)^{L_{A''_3}} = (L'_1)^{L''_1 * L''_2} * (L'_3)^{\tilde{L}''_2 * \tilde{L}''_1}; \\ L_{A'_4-A'_3} = (\tilde{L}'_3)^{L_{A''_3}} * (L'_2)^{L_{A''_2}} = (\tilde{L}'_3)^{\tilde{L}''_2 * \tilde{L}''_1} * (L'_2)^{\tilde{L}''_1 + L''_2} = \\ (\tilde{L}'_3 * L'_2)^{\tilde{L}'_2 * \tilde{L}''_1 + L''_2}. \end{gathered}$$

The knowledge about contexts A'_1, A'_2, A'_3 of the first level is derived as follows:

$$\begin{gathered} L_{A'_1} = (L'_1 * L'_2)^{L''_1 * L''_2 + \tilde{L}''_1} + (L'_1 * L'_3)^{L''_1 * L''_2 + \tilde{L}''_2 * \tilde{L}''_1} = \\ (L'_1 * (L'_2 + L'_3))^{L''_1 * L''_2 + \tilde{L}''_2 * \tilde{L}''_1}; \\ L_{A'_2} = (\tilde{L}'_1)^{L''_1 * L''_2} + (L'_2)^{\tilde{L}''_1 + L''_2} + (L'_3)^{\tilde{L}''_2 * \tilde{L}''_1} = \\ (\tilde{L}'_1 + L'_2 + L'_3)^{L''_1 * L''_2 + \tilde{L}''_2 * \tilde{L}''_1}; \\ L_{A'_3} = (\tilde{L}'_2 * \tilde{L}'_1)^{L''_1 * L''_2 + \tilde{L}''_1} + (\tilde{L}'_2 * L'_3)^{L''_2 + \tilde{L}''_2 * \tilde{L}''_1} = \\ (\tilde{L}'_2 * (\tilde{L}'_1 + L'_3))^{L''_1 * L''_2 + \tilde{L}''_2 * \tilde{L}''_1}. \end{gathered}$$

Now it is possible to derive the interpreted knowledge about the relation between A_1 and A_3 taking all the contexts and metacontexts into account as:

$$\begin{gathered} L_{A_1-A_3} = (L_1)^{L_{A'_1}} * (L_2)^{L_{A'_2}} + (L_3)^{L_{A'_3}} = \\ = (L_1 * L_2 + L_3)^{(L'_1 * L'_2 + L'_1 * L'_3 + \tilde{L}'_2 * \tilde{L}'_1 + \tilde{L}'_2 * L'_3)^{(L''_1 * L''_2 + \tilde{L}''_2 * \tilde{L}''_1)}} \end{gathered}$$

How to interpret formulas of the algebra? The application area defines the way of representing the context and metacontexts. If the internal structure of a context is known, one can, for example, use the formalism of 'semantic balance' between the internal structure of a context and its external relationships [Grebenyuk *et al.*, 1996]. Some examples of an interpretation technique in the domain of temporal relations are given in [Bondarenko *et al.*, 1993]. The contexts of a relationship in [Puuronen and Terziyan, 1992; Terziyan, 1993] are rules that define the conditions when the relations appear and disappear. The knowledge about knowledge sources can also be considered as a context for a knowledge base refinement as in [Puuronen and Terziyan, 1996; Puuronen and Terziyan, 1997b]. Farquhar *et al.* [1995] have used contexts to integrate databases, and Halpern and Moses [1992] have used contexts to reason about knowledge and belief of multiple agents. In

[Puuronen and Terziyan, 1997a], the knowledge about the relationship of an expert with his colleagues is used as a context to interpret knowledge acquired from this expert.

Deriving unknown knowledge that is interpreted when the result of interpretation and the context of interpretation are known (contextualization). This problem occurs when some knowledge has been interpreted in some context and we have all knowledge about this context and the knowledge that is the result of the interpretation. For example, let us suppose that your colleague, whose context you know well, has described you a situation. You use knowledge about the context of this person to interpret the 'real' situation. Example is more complicated if several persons describe you the same situation. In this case, the context of the situation is the semantic sum over all the personal contexts.

This second reasoning problem can be solved using the following equation:

$$L_x^{\langle \text{knowledgeaboutcontext} \rangle} = \langle \text{interpreted knowledge} \rangle.$$

Deriving unknown context of interpretation when the knowledge and its interpretation in this context are known (context recognition). This problem occurs when we have knowledge that has been interpreted in some unknown context and we also know what is the result of the interpretation. For example let us suppose that someone sends you a message describing the situation that you know well. You compare your own knowledge with the knowledge you received. Usually you can derive your opinion about the sender of the message. Knowledge about the source of the message, you derived, can be considered as a certain context in which the real situation has been interpreted and sometimes it can help you to recognize a source or at least his motivation.

This third reasoning problem can be solved using the following equation:

$$\langle \text{knowledge} \rangle^{L_x} = \langle \text{interpreted knowledge} \rangle.$$

Lifting (relative decontextualization). This means deriving knowledge interpreted in some context if it is known how this knowledge was interpreted in another context. This problem is solved by successive solution of the above contextualization and decontextualization problems. Let L_k be the result of the interpretation of some knowledge L_x in the context L_m. The problem is to derive how this knowledge would be interpreted in the context L_n. Thus we have the following procedure of lifting:

$$(L_x^{L_m} = L_k) \Leftrightarrow (L_x = L_k^{\bar{L}_m})[\mathit{contextualization}] \Leftrightarrow$$
$$\Leftrightarrow (L_x^{L_n} = L_k^{\bar{L}_m^{L_n}})[\mathit{decontextualization}].$$

The formal tools, that are necessary to handle these reasoning problems, are presented in Sections 3 and 4 above.

6 CONCLUSION

In our previous papers, we have described metaobjects in a metanetwork as rules, determining the behaviour of relations. In this paper, we propose an interpretation of metaobjects as contexts. This enables the ordering of contexts into a multilevel representation, that can be used during the reasoning process. We have presented a general framework for solving four types of problems: how to derive interpretation of knowledge in a context; how to derive knowledge that was interpreted; how to derive knowledge about the context of an interpretation and how to change knowledge from one context to another.

One way was shown how to interpret expressions of the algebra. Further research is needed to make more universal tools for interpreting such expressions in various application areas where it is reasonable to consider several levels of context. It is also planned to include temporal component to the multilevel knowledge representation by considering dynamically changing contexts. Another important problem is how to use a context if incomplete and inconsistent knowledge about it is acquired from several knowledge sources. It is also necessary to consider cases when inconsistent and incomplete knowledge of multiple experts is interpreted using inconsistent and incomplete knowledge about a context.

Vagan Y. Terziyan
State Technical University of Radioelectronics, Ukraine.

Seppo Puuronen
University of Jyväskylä, Finland.

REFERENCES

[Akman and Surav, 1996] V. Akman and M. Surav. Steps Towards Formalizing Context. *AI Magazine*, **17**, 55–72, 1996

[Attardi and Simi, 1995] G. Attardi and M. Simi. A Formalization of Viewpoints. *Fundamenta Informaticae*, **23**, 149–174, 1995.

[Bondarenko *et al.*, 1993] M. F. Bondarenko, V. A. Grebenyuk and V.Ya. Terziyan. Reasoning Based on the Algebra of Semantic Relation. *Pattern Recognition and Image Analysis*, **3**, 488–499, 1993.

[Brezillon, 1994] P. Brezillon. Context Needs in Cooperative Building of Explanations. In *Proc. of the First European Conference on Cognitive Science in Industry*, pp. 443–450, 1994.

[Brezillon and Abu-Hakima, 1995] P. Brezillon and S. Abu-Hakima. Using Knowledge in its Context: Report on the IJCAI-93 Workshop. *AI Magazine*, **16**, 87–91, 1995.

[Brezillon and Cases, 1995] P. Brezillon and E. Cases. Cooperating for Assisting Intelligently Operators. In *Proc. of Actes International Workshop on the Design of Cooperative Systems*, pp. 370–384, 1995.

[Buvac *et al.*, 1993] S. Buvac, V. Buvac and I. A. Mason. Propositional Logic of Context. In *Proceedings of the Eleventh AAAI Conference*, Washington DC, pp. 412–419, 1993.

[Buvac *et al.*, 1994] S. Buvac, V. Buvac and I. A. Mason. Sematics of Propositional Contexts. In *Proceedings of the Eight International Symposium on Methodologies for Intelligent Systems*, pp. 468–477. Volume 869 of *Lecture Notes in Artificial Intelligence*, Springer Verlag,

[Buvac *et al.*, 1995] S. Buvac, V. Buvac and I. A. Mason. Metamathematics of Contexts. *Fundamenta Informaticae*, **23**, 263–301, 1995.

[Buvac, 1996] S. Buvac. Quantificational Logic of Context. In *Proceedings of the Thirteenth AAAI Conference*, Menlo Park, California, 1996.

[Edmonds, 1997] B. Edmonds. A Simple-Minded Network Model with Context-like Objects. CPM Report 97-15, The workshop on Context at the European Conference on Cognitive Science (ECCS'97), Manchester, April, 1997.

[Farquhar *et al.*, 1995] A. Farquhar, A. Dappert, R. Fikes and W. Pratt. Integrating Information Sources Using Context Logic. Technical Report KSL-95-12, Stanford, 1995.

[Grebenyuk *et al.*, 1996] V. Grebenyuk, H. B. Kaikova, V. Ya. Terziyan and S. Puuronen. The Law of Semantic Balance and its Use in Modeling Possible Worlds. In *STeP-96 - Genes, Nets and Symbols*, pp. 97–103. Publications of the Finnish AI Society, Vaasa, Finland, 1996.

[Guha, 1991] R. V. Guha. *Contexts: A Formalization and Some Applications*, Stanford Ph.D. Thesis, 1991.

[Halpern and Moses, 1992] J. Y. Halpern and Y. Moses. A Guide to Completeness and Complexity for Modal Logic of Knowledge and Belief. *Artificial Intelligence*, **54**, 1992.

[McCarthy, 1993] J. McCarthy. Notes on Formalizing Context. In *Proc. of 13 International Joint Conference on Artificial Intelligence*, pp. 555–560, 1993.

[McCarthy, 1995] J. McCarthy. A Logical AI Approach to Context. Technical Note, Computer Science Department, Stanford University, 1995, Available in http://www-formal.stanford.edu/jmc/index.html.

[Puuronen and Terziyan, 1992] S. Puuronen and V. Ya. Terziyan. A Metasemantic Network. In *New Directions in Artificial Intelligence*, pp. 136–143. Publications of the Finnish AI Society, 1992.

[Puuronen and Terziyan, 1996] S. Puuronen and V.Ya. Terziyan. Modeling Consensus Knowledge from Multiple Sources Based on Semantics of Concepts. In *Challenges of Design, ER'96 International Conference on Conceptual Modeling*, Cottbus, Germany, pp. 133-146, 1996.

[Puuronen and Terziyan, 1997a] S. Puuronen and V. Ya. Terziyan. Colleague-Oriented Interpretation of Knowledge Acquired from Multiple Experts. In *Proceedings of the Joint 1997 Pacific Asian Conference on Expert Systems/Singapore International Conference on Intelligent Systems (PACES/SPICIS 97)*, D. Patterson, G. Leedham, K. Warendorf and T. A. Hwee, eds. pp. 737–741. The Nanyang Technological University, 1997.

[Puuronen and Terziyan, 1997b] S. Puuronen and V.Ya. Terziyan. Voting-Type Technique of the Multiple Expert Knowledge Refinement. In *Proceedings of the Thirtieth Hawaii International Conference on System Sciences*, Vol. V. R. H. Sprague, ed. pp. 287–296. IEEE Computer Society Press, 1997.

[Russel and Wefald, 1991] S. Russel and E. Wefald. Principles of Metareasoning. *Artificial Intelligence*, **49**, 361–395, 1991.

[Shastri, 1989] L. Shastri. Default Reasoning in Semantic Networks: A Formalization of Recognition and Inheritance. *Artificial Intelligence*, **39**, 283–355, 1989.

[Shapiro, 1991] S. C. Shapiro. Cables, Paths and 'Subconscious' Reasoning in Propositional Semantic Networks. In *Principles of Semantic Networks*, J. F. Sowa, ed. pp. 137–156. Morgan Kaufmann, 1991.

[Taylor *et al.*, 1995] W. A. Taylor, D. H. Weimann and P. J. Martin. Knowledge Acquisition and Synthesis in a Multiple Source Multiple Domain Process Context. *Expert Systems with Applications*, **8**, 295-302, 1995.

[Terziyan, 1993] V. Ya. Terziyan. Multilevel Models for Knowledge Bases Control and Their Applications to Automated Information Systems. Post Doctoral Degree Thesis, State Technical University of Radioelectronics, Kharkov, Ukraine, 1993, (in Russian).

[Woods, 1991] W. A. Woods. Understanding Subsumption and Taxonomy: A Framework for Progress. In *Principles of Semantic Networks*, J. F. Sowa, ed. pp. 45–94. Morgan Kaufmann, 1991.

PIERRE BONZON

CONTEXTUAL LEARNING: TOWARDS USING CONTEXTS TO ACHIEVE GENERALITY

1 INTRODUCTION

A difficult task encountered in machine learning, as in many other domains, is to achieve *generality*. Briefly, a solution to a problem is said to be general when it is not bound to data instances describing the problem. In other words, generality allows for the abstraction of solution classes from specific problems. This article presents an attempt to use formalized contexts as a way to achieve generality in machine learning.

To report on this project immediately raises two (at first unrelated) issues, i.e.

- the description of our problem domain (i.e. machine learning)
- our interpretation of the notion of context.

A common goal in machine learning is to synthesize operators (such as operators for controlling robots) in terms of *object-level* or *specific* domain entities (such as robot moves or actions). To illustrate this approach, let us consider a hypothetical trading program with access to the prices of various commodities in different countries. A learning goal for this program could be to synthesize operators representing such rules as 'if the current US market price of crude oil is M then buying a cargo of T tons in Venezuela at price P will yield a net profit $f(M,T,P)$' [Russell, 1989]. Clearly, this solution involves specific entities (such as prices, oil cargoes, buying moves, net profits). This solution for trading oil could hardly be generalized to other trading contexts (which, for instance, do not involve cargoes), not to mention broader, non-trading contexts. Instead, each relevant context could be modelled in terms of decision-theoretic concepts, such as those relating states, actions and utilities, and leading to the compilation of action-utility rules [Russell, 1989]. A greater flexibility could then be obtained by synthesizing operators aggregating *meta-level* operations. As an example of meta-level operations, we shall consider the *inference steps* leading to the discovery of object-level concepts. In our trading example, a meta-level operator based on the lifting of a general opportunity theory could then read (in a simplified way) as follows: '*to instantiate an action-utility pair, first select a lifting axiom by entering any relevant context; then enter the opportunity context and instantiate its theory by substituting lifted values for the action and utility parameters; finally conclude with deductions from the*

P. Bonzon, M. Cavalcanti and R. Nossum (eds.), Formal Aspects of Context, 127–141.

lifted opportunity theory'. In this solution, generic (i.e. context independent) entities have been substituted for specific (i.e. context dependent) ones. As a result, generality has been achieved.

Turning to our interpretation of contexts, they provide us with a declarative representation of both object- and meta-level theories (i.e. sets of formulas) in some well-defined logic. More precisely, our formalized contexts are terms that denote sets of restricted formulas (similar to the Horn clauses of logic programming) from an extended 2-sorted predicate calculus. While this technical description does not constitute a complete or satisfactory interpretation of the notion of context, it does reflect our purely instrumental view of this concept: to us, contexts are syntactic entities that allow for the explicit description and manipulation of other syntactic entities. This fundamental property explains why the use of formalized contexts is crucial in the work reported here: as noted above, our approach to generality in machine learning requires that inference steps be regarded as meta-level operations; as such, they need to be treated as syntactic entities capable of operating on syntactic entities from object-level domains. This property makes our formalized contexts truly reminiscent of other formal systems, such as reflective systems and hierarchical logics.

The formal properties of the computational model resulting from merging contexts and machine learning will not be explored in detail. Our contribution therefore has the flavour of a proposal for initiating work in a new direction. Our aim however is to provide sufficient inside into the modelling of contexts so as to make their role clearly explicit. Likewise, we want to be able to come up with a working implementation of some simple examples. Our emphasis will thus be on formalizing contexts from scratch. These developments do not require any background knowledge on machine learning (the reader interested in acquiring this knowledge is advised to consult Hutchinson's excellent treaty [Hutchinson, 1994]).

The rest of this contribution is organized as follows: in Section 2, we briefly review some basic concepts about reflective systems, hierarchical meta-logics and their relation to formalized contexts; in Section 3, we present a set of context models together with their proof system given under the form of a Prolog procedure; Section 4 introduces our contextual learning model; finally, this model is illustrated with a complete modelization of the example given above.

2 REFLECTIVE SYSTEMS, HIERARCHICAL METALOGICS AND FORMALIZED CONTEXTS: A SHORT REVIEW OF RELATED WORK

According to B. Smith [1984], *reflective systems* consist of an infinite tower of language processors (or interpreters) in which each processor interprets the one below it. The interpreter at the bottom executes user input and an 'ultimate' machine at the top runs the tower. Levels are connected by a mechanism, called *reification*, which converts a program into data and permits a program running at one level to provide code to the next higher level. Conversely, *reflection* activates a level down in the tower and turns data into a program [Jefferson and Friedman, 1992].

The *hierarchical metalogics* proposed by Giunchiglia and Serafini [1994] represent an attempt to 'understand and formalise meta-reasoning as deduction in a logical (declarative) meta-theory'. They follow the work of Weyhrauch [Weyhrauch, 1980], in which the use of a hierarchy of logical meta-theories deductively linked to each other by *reflection rules* was first outlined. As pointed out in [Bowen and Kowalski, 1982], one of these two rules allows the meta-level language to communicate the solutions of its problems to the object language (i.e. reflecting down). The other one allows the object language to communicate the solutions of its problems to the meta-level language (i.e. reflecting up). Building on the theoretical model of hierarchical logics, Giunchiglia and Cimatti [1994] describe a reasoning system which is able 'to introspect its own code, to reason deductively about it in a declarative metatheory and, as a result, to produce new executable code which can then be pushed back into the underlying implementation'. This system therefore extends itself by adding or modifying inference procedures. In contrast, we contemplate a system which extends itself by learning meta-level operators taking the form of useful inference steps.

Although they appear in quite different frameworks, the processes of reflecting up and down in a tower of processors or between an object- and meta-theory are obviously related. In both cases, data is passed between two systems, one of which either operates on or defines properties of the other. Somehow, the introduction of *formalized contexts* can be viewed as an extension of both approaches. When coupled with a *uniform proof system* allowing to enter and leave a context, they either operate on or are related to each other in the same way as the processors in a reflective tower or the languages of hierarchical metalogics.

As outlined above, our formalized contexts can be viewed as terms (of the context sort) that denote sets of formulas (of the discourse sort). Following results on amalgamating object- and meta-level languages into ambivalent logic [Kalsbeek, 1995], we do not introduce names for formulas. A term can thus be used to denote itself in any context. Following the usual presentation of formalized contexts [McCarthy, 1993] (for which a precise description, including a semantics, an Hilbert style proof system and corresponding soundness and completeness can be found in [Buvac *et al.*, 1995; Buvac, 1996]), formulas in our contexts include the usual `ist(C,P)` modality. In this modality, `C` a first-order term of the context sort denoting a context (i.e. standing as its name), `P` is a first order atomic formula of the discourse sort, and `ist(C,P)` is intended to mean that 'sentence `P` holds in context C'. Context descriptions can be either explicit or implicit. Explicit descriptions require assertions of the form $\alpha : \Phi$, where α is a term related to the context and Φ is a set of formulas (see Section 3 for formal definitions). Implicit descriptions of contexts result whenever the term they denote are refereed to in the modality `ist(C,P)`.

3 BASIC MODELS OF CONTEXTS AND THEIR CORRESPONDING PROOF SYSTEM

In what follows, we shall assume a minimal background in first order logic, including the definition of a substitution θ, of an instance $e\theta$ or $E\theta$ (of a single expression e or of a set of expressions E) and of a most general unifier (mgu) of a set of expressions.

3.1 A first model

In our first (introductory) model, an explicit context C is defined by one or more related assertions of the form $\alpha : \Phi$, where α is a term of the context sort that can be unified with C and Φ is a set, and each $\phi \in \Phi$ is either

- an atomic formula of the discourse sort
- an implication $\gamma_1 \rightarrow \gamma_2 \rightarrow \ldots \gamma_n$ where each γ_i is an atomic formula of the discourse sort

and all the variables appearing in either α or Φ are universally quantified.

EXAMPLE:

```
block(s0): [on(a,b), on (b,c)].
block(S) : [on(X,Y) -> above(X,Y)].
```

Throughout this paper, we shall adopt for our examples the PROLOG syntax and conventions. In particular, we use capital letters to represent variables. Following the results of ambivalent logic [Kalsbeek, 1995], we do not distinguish between ordinary and meta variables, the former, such as C in `ist(C,P)`, standing for terms, the later (such as P) standing for arbitrary formulas.

Thus, s0 being a situation constant, the first assertion above relates to (and partially defines) the context of a block world in a given situation s0; in contrast, S is a situation variable and the second assertion relates to a block world context in any situation, thus defining a rule applicable in any block world context.

In this simple framework, a proof system for deriving and/or instantiating `ist(C,P)` modalities is given by a Prolog procedure which is equivalent to the standard meta-interpreter for Prolog itself, i.e.

```
ist(C,Q):-instance(C,Q);
          instance(C,P->Q),
          ist(C,P).
```

with

```
instance (C,P):- C:L, member(P,L).
```

where ";" and "," represent a disjunction and a conjunction respectively, and `instance(C,P)` retrieves from the assertions $\alpha : \Phi$ related to `C` (i.e. such that $C\theta = \alpha\theta$, where θ is a most general unifier of `C` and α) and each and every formula $\phi\Theta \in \Phi\theta$ that can be unified with `P`.

EXAMPLE.
Given the above two assertions and the query

```
?- ist(block(S),above(a,Y)).
```

the proof system will first come up with

```
instance(block(s0), on(a,Y)-> above(a,Y))
```

followed, through the recursive call `ist(C,P)`, by

```
instance(block(s0), on(a,b))
```

and thus will instantiate the query to

```
ist(block(s0), above(a,b)).
```

3.2 Models of nested contexts

In our next (more general) model, the formulas γ_i that appear in the implications $\phi \in \Phi$ can be themselves `ist` modalities. Furthermore, these modalities can be possibly nested, i.e. of the form `ist(C0,ist(C1,..., ist(Cn,P)))`, and/or contain implications and meta-variables standing for arbitrary formulas. Formally, the language **L** of all possible formulas $\phi \in \Phi$ can be defined as the least set satisfying the equation $\mathbf{L} = \mathbf{A} \cup \mathbf{L} \rightarrow \mathbf{L} \cup \mathtt{ist}(\mathbf{C}, \mathbf{L})$, where **A** is the set of atomic formulas of the discourse sort and **C** the set of terms of the context sort.

EXAMPLE.

```
c(0): [ist(block(s0), on(a,b)),
       ist(block(S), on(X,Y)->above(X,Y)),
       ist(C,P->Q)-> ist(C,P)-> ist(C,Q)].
```

Depending on their location, implications have different intuitive interpretations. Modalities containing implications, such as

```
ist(block(S), on(X,Y)-> above(X,Y))
```

simply express (in the above example, within `c(0)` considered as an explicit *outer* context) that an implication holds in some implicit *inner* contexts `block(S)`. Implications involving modalities, such as,

```
ist(C,P->Q)-> ist(C,P)-> ist(C,Q)
```

are called *meta-implications* (the `ist` modality is then interpreted as a meta-level predicate) representing inference rules (here,*modus ponens*) applicable in their outer context.

N.B. Successive implications associate to the right, i.e. `A->B->C = A->(B->C)`.

An extended proof procedure for nested modalities can be obtained from the standard meta-interpreter given above by simply replacing the call `instance(C,P->Q)` by an additional recursive call `ist(C,P->Q)`, i.e.

```
ist(C,Q):- instance(C,Q);
           ist(C,P->Q),
           ist(C,P).
```

Reflecting the interpretation given above, this extended procedure does not directly use implications contained within modalities. It relies instead on inference rules given by meta-implications. Towards that goal, successive recursive calls `ist(C,P->Q)` will *unroll* meta-implications associating to the right.

EXAMPLE.

Consider the nested query

(1) $? - \texttt{ist}(\texttt{c}(0), \texttt{ist}(\texttt{block}(\texttt{S}), \texttt{above}(\texttt{a}, \texttt{Y})))$.

In this particular case, modus ponens will be put to work through a double recursive call of `ist(C,P->Q)`. (1) will first lead to the recursive calls

(2) $\texttt{ist}(\texttt{c}(0), \texttt{P}' - > \texttt{ist}(\texttt{block}(\texttt{S}), \texttt{above}(\texttt{a}, \texttt{Y})))$,

(3) $\texttt{ist}(\texttt{c}(0), \texttt{P}')$.

(2) will then lead to the recursive calls

(4) $\texttt{ist}(\texttt{c}(0), \texttt{P}'' - > \texttt{P}' - > \texttt{ist}(\texttt{block}(\texttt{S}), \texttt{above}(\texttt{a}, \texttt{Y})))$

(5) $\texttt{ist}(\texttt{c}(0), \texttt{P}'')$.

(4) in turn will give rise to the instantiation

```
ist(c(0), ist(block(S),P''' -> above(a,Y))
          -> ist(block(S),P''')
               -> ist(block(S),above(a,Y)))
```

As a result, (5) will be instantiated as

```
ist(c(0), ist(block(S), on(a,Y) -> above(a,Y))).
```

Finally, (3) will become

```
ist(c(0), ist(block(s0),on(a,b)))
```

thus allowing for the instantiation of (1) as

```
ist(c(0), ist(block(s0),above(a,b))).
```

3.3 A model of explicit object- and meta-level contexts

In our previous model, the outer context, i.e. `c(0)`, was explicit, but the inner ones were left implicit. Our next model allows for the explicit definition of both inner and outer contexts. When they contain only object- and meta-level formulas respectively, they will be called ***object-*** and ***meta-level contexts***. In the example below, the `block(_)` are thus object-level contexts and `c(0)` is a meta-level context:

```
block(s0):[on(a,b), on(b,c)].
block(S) :[on(X,Y) -> above(X,Y)].
c(0):[ist(C,P->Q) -> ist(C,P) -> ist(C,Q)].
```

In order to access object-level contexts from within meta-level contexts, the proof procedure must extended with an additional clause, i.e.

```
ist(C1,ist(C,P)):- ist(C,P).
```

The additional clause, which reflects a property of contexts called *semi-flatness,* accounts for the possibility to *enter* and *leave* contexts. As suggested in [McCarthy, 1993], these two operations are analogous to making and subsequently discharging ***assumptions*** in natural deduction systems. Our extended proof procedure augmented with semi-flatness has been proved equivalent to a natural deduction system using explicit context assertions [Bonzon, 1997].

N.B. In many cases, the various recursive calls in our proof procedure will lead to an infinite recursion. In order to get an effective automated proof system one may implement an ***iterative deepening search*** as follows:

```
search(ist(C,Q)/N) :-  ist(C,Q)/N;
                       N1 is N+1,
                       search(ist(C,Q)/N1).
```

where

```
ist(C,Q)/N :- instance(C,Q);
              N>0,
              N1 is N-1,
              ist(C,P->Q)/N1,
              ist(C,P)/N1.

ist(C0,ist(C,P))/N :- ist(C,P)/N.
```

and with the top level call for any nested `ist(C,Q)` defined as

```
ist(C,Q) :- search(ist(C,Q)/1).
```

3.4 *A model involving the lifting of theories*

Our next model elaborates on McCarthy's original proposal [McCarthy, 1993] for lifting theories in contexts. Theories (i.e. rules applicable in various contexts) are now stated in their own context assertions (e.g. as in `above_theory` below). In order to access these theories, selected object-level contexts must be extended with *lifting axioms*. Furthermore, to enforce these lifting axioms, meta-level contexts must include a new *lifting inference rule*. Our standard example looks as follows:

```
above_theory:[on(X,Y)-> above(X,Y)].
block(s0):[on(a,b)].
block(S) :[ist(above_theory,P)
            -> ist(block(S),P)].

c(0):[ist(C,P->Q) -> ist(C,P) -> ist(C,Q),
      ist(C, ist(A,P) -> ist(C,P))
       -> ist(A,P)
        -> ist(C,P)].
```

According to the lifting axiom asserted within `block(S)`, any sentence *P* stated within `above_theory` also holds within any `block(S)`. According to the lifting inference rule asserted within `c(0)`, if the condition `ist(A,P)` of a lifting axiom stated within context `C` holds, then the sentence `P` holds within `C`. It should be noted that in contrast to [McCarthy, 1993] and as postulated in [Attardi and Simi, 1994], our implementation of theory lifting relies on an explicit outer context. However, in contrast to [Attardi and Simi, 1994], whose lifting axiom is bound to the `above_theory` context, our lifting inference rule is general.

3.5 *Reflective contexts*

We define *reflective contexts* as contexts that keep track of inference steps when deriving `ist(C,P)` modalities. They involve a new kind of modalities `reflect(V,W)` meaning that modality `W` holds because of the sequence of inference steps `V`. In order to relate to modalities, inference steps must be reified, i.e. represented as data structures taking the form of deduction *traces*. As a simple example, let us consider the following object- and meta-level contexts (which do not include theory lifting):

```
block(s0):[on(a,b),
           on(X,Y) -> above(X,Y)].
c(0):[ist(C,P->Q)  -> ist(C,P) -> ist(C,Q)].
```

The corresponding reflective context $r(0)$ is as follows:

```
r(0):[instance(C,P) -> reflect(axiom(C),ist(C,P)),
      reflect(X,ist(C,P->Q))
        -> reflect(Y,ist(C,P))
          -> reflect(mp(X,Y),ist(C,Q))].
```

Its first rule reflects an *axiom* instantiation (i.e. instantiating a fact related to a given context), as performed by the 'ultimate' machine (i.e. the proof system calling the `instance(C,P)` procedure).

N.B. As `instance(C,P)` is actually a Prolog procedure (defined in 3.1), in order for such a call to appear in a context assertion, the proof system must be extended with an additional clause reflecting semi-flatness properties, i.e.

```
ist(C1,instance(C,P)):- instance(C,P).
```

The second rule reflects an application of *modus ponens* as defined in `c(0)`. It reifies this application through the trace `mp(X,Y)`, where `X` and `Y` are the reified inference steps needed to deduce the antecedents `P->Q` and `P`, respectively.

As an example, consider the following call and its resulting instantiations for the trace `D` and goal predicate `ist(C,above(a,Y))`:

```
?- ist(r(0),reflect(D,ist(C,above(a,Y)))).
D = mp(axiom(block(s0)),axiom(block(s0))) ,
C = block(s0),
Y = b.
```

The *fully instantiated* trace `D` shows the inference steps taken during this deduction (i.e. an application of modus ponens with both antecedents resulting from an axiom instantiation from the `block(s0)` context). Conversely, this fully instantiated trace `D` can be *forced* back into `r(0)`, by using the same call as above with D fully instantiated: this will drive the deduction into the same inference step and produce the same instantiations for the goal predicate. Driving a deduction by forcing a fully instantiated trace leads to a reduced search space. This makes `D` equivalent to a meta-level operator for deducing `above(X,Y)` in context `block(s0)`. A *partially* instantiated trace, such as

```
D = mp(axiom(block(_)),axiom(block(_)))
```

forced into the same `r(0)` would similarly (though to a lesser extend) reduce the search space and produce the same instantiations for the goal predicate. This would make it equivalent to a meta-level operator for deducing `above(X,Y)` in any context `block(S)`.

4 APPLICATION: A CONTEXTUAL LEARNING MODEL BASED ON THEORY LIFTING

As an application of the preceding developments, let us now consider a contextual learning model based on theory lifting. Following a traditional approach in machine learning (see for instance [Hutchinson, 1994]), this particular learning model will be defined in terms of *inputs* and *goals*. Inputs will be given under the form of various contexts assertion, i.e.

- a collection of object-level theories defining application domains, e.g. `block(s0)`
- a collection of *liftable* object-level theories together with *lifting axioms*, e.g. `above_theory` and `block(S)` of Section 3.4.

The goal of the system is to discover *facts* (e.g. concept definitions, operators, rules, or simply instantiations of a goal predicate) which follow from the lifting of theories into selected object-level contexts.

As outlined in the introduction, our aim is to produce general meta-level operators leading to the discovery of such facts. As we have just seen, deduction traces forced back into reflective contexts constitute an example of such meta-level operators. In our particular case, meta-level operators will be given under the form of *partially* instantiated traces. As the corresponding inference steps are not bound to specific data instances, these operators will be truly general.

In order to derive theses traces, the system will first hop up to a top meta-level reflective context, thus accounting for the *learning* of a meta-level operator. Hopping down again to force these traces will then be equivalent to *applying* this operator. The following additional context assertions are thus needed to implement this learning scheme:

- a top meta-level context `r(1)` allowing for the derivation of partially instantiated traces
- a meta-level context `r(0)` to be forced by partially instantiated traces (to get complete instantiations).

A possible choice for partially instantiated traces is to specify only the first antecedent (i.e. an implication) of each modus ponens. This process is equivalent to a kind of *partial deduction* with respect to theory lifting. The corresponding top meta-level context r(1) associates a reflective form of the lifting inference rule of Section 3.4 with a shortened form of reflective modus ponens, i.e.

```
r(1):[instance(C,ist(A,P) -> ist(C,P))
      ->instance(A,P)
         ->reflect(lift(axiom(C),axiom(A)),ist(C,P)),
      reflect(X,ist(C,P->Q))
       ->reflect(mp(X,_),ist(C,Q))].
```

As an example, the call

```
ist(r(1),reflect(D,ist(C,above(X,Y))))
```

will lead to the following partial instantiation for D

```
D =   mp(lift(axiom(block(_)),axiom(above_theory)),_).
```

As this trace clearly shows, the inference steps taken during this partial deduction consist of an application of modus ponens whose first antecedent follows from lifting above_theory into any context block(_). The corresponding meta-level operator can be seen as an instantiation of the following control rule:

> in order to get a fact, first select a lifting axiom; then instantiate a liftable theory; finally conclude by applying modus ponens to an implication from the lifted theory.

While this control rule itself was not known a priori, an appropriate partial instantiation has been *learned* under the form of the partially instantiated trace D.

Finally, a meta-level context r(0) allowing for the complete instantiation of the goal predicate could be

```
r(0): [instance(C,P) ->reflect(axiom(C),ist(C,P)),
       instance(C,ist(A,P) -> ist(C,P))
         -> instance(A,P)
              -> reflect(lift(axiom(C),axiom(A)), ist(C,P)),

       reflect(X,ist(C,P->Q)),
         -> reflect(Y,ist(C,P))
              -> reflect(mp(X,Y),ist(C,Q))].
```

With regard to our previous r(0) given in Section 3.5, this context assertion further includes the reflective version of the lifting inference rule just introduced in r(1).

5 EXTENDED EXAMPLE: THE COMPILATION OF HETEROGENEOUS KNOWLEDGE

We conclude with an illustration providing a complete modelization of our introductory example (elaborated after [Russell, 1989]). Suppose that our hypothetical trading program has access to data indicating that

- cargo-load of 100 000 tons of oil is offered for sale in Venezuela
- the price of oil in Venezuela and in the US are $10 and $12 a ton, respectively
- the cost of shipping oil from Venezuela to the US is $0.5 a ton.

From this data, one can easily deduce that buying these 100,000 tons of Venezuelan oil to ship and sell them in the US would lead to a profit of $150,000. In order to have a system discover such an opportunity, it was advocated in [Russell, 1989] that a trading program could be instructed to learn such rules as 'if the current US market price of crude oil is M then buying a cargo of T tons in Venezuela at price P will yield a net profit $f(M,T,P)$' for some known f. We propose to come up instead with a meta-level operator applicable in various contexts.

Toward this end, we propose to model relevant contexts by using different instances of decision-theoretic concepts for deriving action-utility rules [Russell, 1989]. In the case of trading oil, this leads us first to the following context assertion, where `contract` and `profit` are instances of the *action* and *utility* concepts, respectively:

```
trade(oil): [cargo(venezuela,100000),
             price(venezuela,10),
             price(us,12),
             shipping(venezuela,us,0.5),
             (cargo(O,T),
              price(O,P)) -> contract(T,P,O),
             (contract(T,P,O),
              price(D,M),
              shipping(O,D,S),
              F is T*(M-P-S),
              F>0)   -> profit(F,D)].
```

We then need a liftable *opportunity* theory for compiling and applying action-utility rules. Towards this end, let us consider the following context assertion, where the Prolog syntax `(|X)` represents a variable length argument list:

```
opportunity(Action,Utility): [(ActionCond -> Action(|X),
                               UtilityCond -> Utility(|Y),
                               compute(ActionCond -> Action(|X),
```

```
                                    UtilityCond -> Utility(|Y),
                                OpportunityCond))
                            -> OpportunityCond
                             -> pair(Action(|X),Utility(|Y))].
```

The operations specified within this theory can be explained as follows: rules related to possible *actions* and *utilities* are first retrieved; if such rules exist, they are used to compute *opportunity* conditions; finally, these conditions are applied to instantiate *action-utility pairs*. More specifically, in this particular case, opportunity conditions correspond to the partial evaluation of a given utility with respect to an applicable action. We thus have the following definition:

```
compute(C->P,D->Q,R) :-  intersection(P,D,Intersection),
                         difference(D,Intersection,Difference),
                         union(C,Difference,R).
```

where **intersection**, **difference** and **union** are the usual operations on sets.
In order to lift this theory into the trade context, let us follow Section 3.4:

```
trade(S):  [ist(opportunity(contract,profit),P)
              -> ist(trade(S),P)].
```

In addition, the iterative proof system of Section 3.3 must be extended to include conjunctive and system calls (such as those involving arithmetic computations):

```
ist(C,(P,Q))/N  :-  ist(C,P)/N,
                    ist(C,Q)/N.

ist(C,P)/N   :- system(P),call(P).
```

The query

```
ist(r(1),reflect(D,ist(C,pair(A,B))))
```

will then lead to the following partial instantiations, with D being equivalent to a meta-level operator for deriving action-utility pair(A,B) in any trade context:

```
D = mp(mp(lift(axiom(trade(_)),
                     axiom(opportunity(contract,profit))), _),_) ,
C = trade(_) ,
A = contract(|_),
B = profit(|_).
```

As reflected in D, two embedded modus ponens applications and a lifting step are required in order to get a partial instantiation for pair(A,B). In order to get a complete instantiation (taking into account the second antecedent of each modus ponens application), D can be forced into r(0). Hopping up first to r(1) and then down to r(0) defines a new outer context assertion for solving problems

```
solve: [(ist(r(1),reflect(D,ist(C,P))),
         ist(r(0),reflect(D,ist(C,P))))
          -> ist(C,P)].
```

A query involving both the *learning* and the *application* of a meta-level operator and thus leading to a complete instantiation for pair(A,B)) is then

```
?- ist(solve,ist(C,pair(A,B))).
C= trade(oil),
A= contract(100000,10,venezuela),
B= profit(150000,us).
```

The partial deduction that was performed using r(1) could be seen as an application of the following control rule:

> to find an opportunity taking the form of an action-utility pair, first select a lifting axiom by entering any trade context; enter the opportunity context and instantiate its theory by substituting the value contract and profit for the Action and Utility parameters; using an implication from the lifted theory as first antecedent, apply modus ponens to obtain opportunity conditions; finally conclude by applying modus ponens using the opportunity conditions as first antecedent.

While this rather complex control rule itself was not known a priori, it has been first learned in r(1) under the form of a partially instantiated trace D. It was then made available in r(0) for application in all relevant contexts. As a result, the compilation of heterogeneous knowledge into a general meta-level operator has been achieved.

6 CONCLUSIONS

We first presented a meta-level inference architecture for contextual deduction. Treating contexts as syntactic constructs, it allows for the explicit representation of both object- and meta-level theories. A uniform proof system (represented by a modified interpreter for logic programs) allows one to hop up and down the hierarchy of contexts. An iterative deepening search of the corresponding state space prevents infinite recursion and ensures successful termination whenever possible. The resulting computational system resembles very much the tower architecture defined for functional programming, in which each level represents a meta-level operating on the preceding one.

On this basis, we developed a contextual learning model where sequences of inference steps leading to the discovery of object-level concepts can be stored for later reuse. While it is notoriously difficult to abstract classes of operators from specific facts using classical learning models, the inference steps for lifting theories in contexts are not bound to data instances. This overall approach thus represents a tentative way to directly achieve *generality* in learning. More work is needed however to appreciate (eventually) the potential of this new approach.

University of Lausanne, Switzerland.

REFERENCES

[Attardi and Simi, 1994] G. Attardi and M. Simi. Proofs in Contexts. In *Proceedings 4th International Conference on Principles of Knowledge Representation and Reasoning* (KR 94), 1994.

[Bonzon, 1997] P. Bonzon. A Reflexive Proof System for Reasoning in Contexts. In *Proceedings 14th National Conference on Artificial Intelligence* (AAAI 97), 1997.

[Bowen and Kowalski, 1982] K. Bowen and R. Kowalski, Amalgamating Language and Metalanguage in Logic Programming. In *Logic Programming*, K. Clark and S. Tarnlund (eds), Academic Press, 1982.

[Buvac *et al.*, 1995] S. Buvac, V. Buvac and I. A. Mason. Metamathematics of contexts, *Fundamenta Informaticae*, 23 (3), 1995.

[Buvac, 1996] S. Buvac. Quantificational Logic of Context. In *Proceedings 13th National Conference on Artificial Intelligence* (AAAI96), 1996.

[Giunchiglia and Serafini, 1994] F. Giunchiglia and L. Serafini. Multilanguage hierarchical logics, or: how can we do without modal logics, *Artificial Intelligence*, 65, 29–70, 1994.

[Giunchiglia and Cimatti, 1994] F. Giunchiglia and A. Cimatti. Introspective Metatheoretic Reasoning. In *Proc. 4th Internat. Workshop on Meta Programming in Logic* (META 94), 1994.

[Hutchinson, 1994] A. Hutchinson. *Algorithmic Learning*, Clarendon Press, Oxford, 1994.

[Jefferson and Friedman, 1992] S. Jefferson and D. Friedman. A Simple Reflective Interpreter. InY.Yonezawa and B. Smith (eds.), *Reflection and Meta-Level Architecture*, Proc. IMSA Workshop, 1992

[Kalsbeek, 1995] M. Kalsbeek. Correctness of the Vanilla Meta-Interpreter and Ambivalent Syntax. In K. Apt and F. Turini (eds.), *Meta-Logics and Logic Programming*, MIT Press, 1995.

[McCarthy, 1993] J. McCarthy. Notes on Formalizing Context. In *Proceedings. 13th International Joint Conference on Artificial Intelligence* (IJCAI93), 1993.

[Russell, 1989] S. Russell. Execution Architectures and Compilation. In *Proceedings 11th International Join Conference on Artificial Intelligence* (IJCAI89), 1989.

[Smith, 1984] B. C. Smith. Reflection and Semantics in Lisp, Conference Records. In *Proceedings of the 11th ACM Symposium on Principles of Programming Languages* (POPL 84), 1984.

[Weyhrauch, 1980] R. Weyhrauch. Prolegomena to a Theory of Mechanized Reasoning, *Artificial Intelligence*, 13 (1), 133–176, 1980.

LEENDERT W. N. VAN DER TORRE AND YAO-HUA TAN

CONTEXTUAL DEONTIC LOGIC: VIOLATION CONTEXTS AND FACTUAL DEFEASIBILITY

1 INTRODUCTION

It is well-known in deontic logic that there are striking similarities between deontic reasoning and contextual reasoning. For example, in temporal deontic reasoning Thomason [1981] makes a distinction between the context of deliberation and the context of justification to distinguish a (deliberative) ought that implies 'practical-temporal can' from a (judgmental) ought that does not imply 'practical-temporal can.' The truth values of deontic sentences are not only time-dependent in the same, familiar way that the truth values of all tensed sentences are time-dependent, but they are also dependent of a set of choices or future options that varies as a function of time. The context of deliberation is the set of choices when you are looking for practical advice, whereas the context of justification is the set of choices for someone who is judging you. The following example illustrates the distinction between the two contexts.

EXAMPLE 1. Yesterday (t_1) I had both the judgmental and deliberative obligation to go tonight (t_2) to my father's birthday in Rotterdam, so at t_1 it holds that $O_j f \wedge O_d f$. However, yesterday I took the airplane to Brazil, and today I therefore no longer have the deliberative obligation to go to my father's birthday. At t_2 it holds that $O_j f \wedge \neg O_d f \wedge \neg f$. Consequently, now I cannot be in time any more for my father's birthday and I have violated my judgmental ought, but not my deliberative ought.

Deontic logic is hampered by the so-called deontic paradoxes. The contrary-to-duty paradoxes like the notorious Chisholm paradox are the classic benchmark problems of deontic logics, which have initiated developments of monadic deontic logics [Chisholm, 1963; Forrester, 1984], dyadic deontic logics [Tomberlin, 1981] and temporal deontic logics [van Eck, 1982; van der Torre & Tan, 1998b]. Contexts can be distinguished in several deontic paradoxes, for example in Castañeda's paradox of the second best plan [Castañeda, 1981; Feldmann, 1990; Yu, 1995].

EXAMPLE 2 (Second best plan paradox). Suppose dr. Denton is going to give treatment to one of his patients. His best course of action would be to give an aspirin today and an aspirin tomorrow. His second best plan would be to give a Buffering today and another Buffering tomorrow. It would be very bad for him to give mixed medications. If dr. Denton does not give an aspirin today, then there are two contradictory obligations. *'The doctor ought to give an aspirin today and an aspirin tomorrow'* refers to the best plan context, and *'he ought not to give an aspirin tomorrow'* refers to the second best plan context.

P. Bonzon, M. Cavalcanti and R. Nossum (eds.), Formal Aspects of Context, 143–160.

In *a-temporal* deontic reasoning Prakken and Sergot [1996] distinguish between ideal and varying sub-ideal contexts to formalize contrary-to-duty reasoning. Obligations that refer to the ideal context are called ideal (or primary) obligations, and obligations that refer to the sub-ideal context are called contrary-to-duty (or secondary) obligations. Obviously there are relations between the distinction between justification and deliberation on the one hand, and the distinction between ideal and sub-ideal on the other hand. The essential contrast in both distinctions is a difference in what we take as being settled [Hansson, 1971; Thomason, 1981; Loewer & Belzer, 1983]. The following example illustrates the distinction between the ideal and sub-ideal context by Prakken and Sergot's cottage housing regulations, in propositional modal logic alphabetic variants of the notorious Forrester and Chisholm paradoxes [Forrester, 1984; Chisholm, 1963].

EXAMPLE 3 (Cottage housing regulations). First, consider the two obligations *'there ought to be no fence'* and *'if there is a fence, then it ought to be white.'* If there is a fence around the cottage, then there is an obligation *'there ought to be no fence'* referring to the ideal context there is no fence, and there is an obligation *'there ought to be a white fence'* referring to the sub-ideal context there is a fence. Removing the fence as well as painting it white are improvements of the present state of affairs, and therefore obligatory. Whether the fence ought to be removed or painted depends on the context referred to. The two derived obligations *'there ought to be no fence'* and *'there ought to be a white fence'* are contradictory. In a logic that does not distinguish between contexts, like Standard Deontic Logic, the two sentences of the regulations are inconsistent with the fact *'there is a fence.'*

Moreover, consider the three obligations *'there ought not to be a dog'*, *'if there is no dog, then there ought to be no sign'* and *'if there is a dog, then there ought to be a sign'*. If there is a dog, then the obligation *'there ought to be no sign'* is in force in the ideal context there is no dog nor a sign,[1] but it is not in force in the sub-ideal context there is a dog. Moreover, the two derived obligations *'there ought to be no sign'* (ideal context) and *'there ought to be a sign'* (sub-ideal context) are contradictory. Again, the three sentences of the regulations are inconsistent with the fact *'there is a dog'* in, for example, Standard Deontic Logic.

In this paper we introduce Contextual Deontic Logic (CDL), in which we write $O_\gamma(\alpha|\beta)$ for 'α ought to be (done) if β is (done) in the context where γ is (done).' Thus, we represent the context by a propositional sentence (as in [Prakken & Sergot, 1996]). There are several different ways in which $O_\gamma(\alpha\,|\,\beta)$ can be defined. The logic proposed in this paper does *not* formalize prima facie obligations, i.e. obligations *cannot* be overridden by other, stronger obligations, but it consistently formalizes contrary-to-duty reasoning as it occurs in the cottage housing regula-

[1] Contexts are not only useful to analyze contrary-to-duty reasoning, but they can also be used to analyze according-to-duty reasoning related to so-called deontic detachment, also called deontic transitivity. Deontic detachment is notorious for its role in the Chisholm paradox, where it is combined with contrary-to-duty reasoning. However, as Prakken and Sergot [1996] observe, the problem of the Chisholm paradox and deontic detachment is different from the problem of formalizing contrary-to-duty reasoning.

tions. The crucial observation formalized by the logic is that the optimal state, and therefore the obligations, can change radically when the violation context changes. In such cases we say that obligations only in force in the previous violation context are defeated; contextual deontic logic is therefore a defeasible deontic logic. We write $O(\alpha \mid \beta \backslash \neg\gamma)$ for 'α ought to be (done) if β is (done) unless $\neg\gamma$ is (done).' The unless clause can be compared to the justification in a Reiter default 'α is normally the case if β is the case unless $\neg\gamma$ is the case,' written as $\beta : \gamma/\alpha$ [Reiter, 1980]. For example, *'birds fly unless they are penguins'* can be represented by $b : \neg p/f$, and *'penguins do not fly'* by $(b \wedge p) : \top/\neg f$. Hence, the unless clause formalizes a consistency check. The contextual obligations are defined in terms of the defeasible obligations as follows.

$$O_\gamma(\alpha|\beta) =_{def} O(\alpha|\beta\backslash\neg\gamma)$$

Consequently, we disagree with Prakken and Sergot that "contrary-to-duty is not an instance of defeasible reasoning, and that methods of non-monotonic reasoning are inadequate since they are unable to distinguish between defeasibility and violation of primary obligations" [Prakken & Sergot, 1996].

- First, it follows directly from the definition above that contextual reasoning is defeasible, and given that "a contrary-to-duty obligation pertains to, or presupposes, a certain context in which a primary obligation is already violated," contrary-to-duty reasoning is defeasible too! Obligations lack unrestricted strengthening of the antecedent, the typical property of defeasible conditionals [Alchourrón, 1993], to block the combination of obligations from different contexts. Combining obligations of different contexts is counterintuitive, whether these obligations conflict (as in Example 3) or not (called pragmatic oddities in [Prakken & Sergot, 1996][2]).

- Second, the diagnostic problem of *defeasible deontic logic* – i.e. to determine whether 'α ought to be (done) and $\neg\alpha$ is (done)' is a violation or an exception – can be solved by methods of non-monotonic reasoning. In the proof theory 'defeasibility' (in Prakken and Sergot's sense) and violation of primary obligations can be distinguished by distinguishing between different types of defeasibility (see below), and in the semantics they can be distinguished by multi-preference semantics for the strong overridden defeasibility of defeasible obligations 'normally, α ought to be (done)' and by strengths of obligations (priorities) for the weak overridden defeasibility of prima facie obligations [van der Torre & Tan, 1995]; [1997; 1998c].

[2]It is counterintuitive to derive the obligation *'you should keep a promise and apologize for not keeping it'* from the two obligations *'you should keep a promise'* and *'if you do not keep a promise, then you should apologize for it'* and the fact *'you break a promise.'* The two obligations refer respectively to the ideal context you keep your promise and to the sub-ideal context you do not keep your promise, and therefore they should not be combined.

We use CDL to analyze the relation between deontic, contextual and defeasible reasoning. For this analysis, we distinguish between three different types of defeasibility in defeasible reasoning: factual defeasibility (FD), conflict defeasibility (CD) and overridden defeasibility (OD). FD is the only one that does not depend on conflicts between formulas (when written as preferences, as explained later in this paper). The distinction between CD and OD is that in the last case the conflict is resolved, for example by the specificity principle or by priorities. All three types of defeasibility lead to restrictions on strengthening of the antecedent, but they can be discriminated with respect to non-monotonicity in (extensions of) dyadic logic without factual detachment. In such a logic OD leads to non-monotonicity whereas FD does not. Moreover, whether CD gives rise to non-monotonicity depends on the philosophical stance towards dilemmas.

In our view the relation between deontic, contextual and defeasible reasoning is as follows. Deontic contrary-to-duty reasoning is a kind of contextual reasoning, and it is therefore defeasible. The pragmatic oddities show that non-monotonic 'consistency restoring' techniques are insufficient to formalize contrary-to-duty reasoning [van der Torre & Tan, 1995]; [1997], although the inconsistency of the two sets of obligations of the cottage housing regulations in Example 3 in SDL might suggest otherwise [Ryu & Lee, 1993; McCarty, 1994; Yu, 1995]. Consequently, CDL should not be based on OD. It has to be based on factual defeasibility and it is therefore monotonic.[3]

This paper is organized as follows. In Section 2 we introduce contextual obligations $O_\gamma(\alpha \mid \beta)$ and $O(\alpha \mid \beta \backslash \gamma)$ in a preference-based logic and in Section 3 we axiomatize contextual deontic logic in a *modal* preference logic. In Example 4 we analyze contrary-to-duty reasoning in CDL and in Section 5 we discuss the relation between obligations and different types of defeasibility.

2 PREFERENCE-BASED CONTEXTUAL DEONTIC REASONING

We restrict our analysis to *preference-based* contextual deontic logic. We propose the following simple formalization of contextual deontic reasoning, where $\alpha_1 \succ \alpha_2$ is read as 'α_1 is deontically preferred to α_2' or as 'α_1 is better than α_2.'

$$O_\gamma(\alpha|\beta) =_{def} O(\alpha|\beta\backslash\neg\gamma) =_{def} \alpha \wedge \beta \wedge \gamma \succ \neg\alpha \wedge \beta$$

This is a straightforward generalization of preference-based deontic logics defined by $O(\alpha \mid \beta) =_{def} \alpha \wedge \beta \succ \neg\alpha \wedge \beta$, see e.g. [van der Torre, 1997]. In fact,

[3]At first sight, the monotonicity property of CDL may seem surprising. However, this property is a consequence of the restricted language, and the lack of factual detachment. If a notion of factual detachment would be added, then the logic would become non-monotonic. Hence, we do use non-monotonic techniques! This surprising combination of defeasibility and monotonicity can be further explained by comparing contextual deontic logic with Reiter defaults, which are also monotonic. Non-monotonicity only arises in Reiter's default logic when default rules are applied, which is the default logic analogue of factual detachment, to construct so-called Reiter extensions.

preference-based dyadic obligations can be defined as a special case of preference-based contextual obligations.

$$O(\alpha|\beta) =_{def} O_{\top}(\alpha|\beta)$$

Despite their advantages, preference-based deontic logics also validate the theorem $O(\alpha \mid \beta) \leftrightarrow O(\alpha \wedge \beta \mid \beta)$, which is counterintuitive at first sight. For a discussion on this theorem, see e.g. [Hansson, 1971]. Moreover, it is easily seen that preference-based contextual deontic logic has a similar theorem for the context γ, namely $O_{\gamma}(\alpha \mid \beta) \leftrightarrow O_{\gamma \wedge \beta \wedge \alpha}(\alpha \mid \beta)$. These two theorems, characteristic for preference-based logics, are not further discussed in this paper, because we are interested in the relation between deontic, contextual and defeasible reasoning.

In this paper we only consider strong preferences, i.e. preferences that have the following property of restricted (left and right) strengthening, that formalizes that there is no overridden defeasibility in the logic. Strengthening is restricted by a 'consistency check' that formalizes that 'ought implies can,' in which $\diamondsuit \alpha$ stand for 'α is possible' or 'α is propositionally consistent.'

$$(\alpha_1 \succ \beta_1) \wedge \diamondsuit(\alpha_1 \wedge \alpha_2) \rightarrow (\alpha_1 \wedge \alpha_2 \succ \beta_1 \wedge \beta_2)$$

This restricted strengthening of the preferences, i.e. lack of overridden defeasibility, corresponds for the contextual obligations to restricted strengthening of the antecedent as well as restricted strengthening of the context.

$$\frac{O_{\gamma}(\alpha|\beta), \diamondsuit(\alpha \wedge \beta \wedge \beta' \wedge \gamma)}{O_{\gamma}(\alpha|\beta \wedge \beta')} \qquad \frac{O(\alpha|\beta\backslash\neg\gamma), \diamondsuit(\alpha \wedge \beta \wedge \beta' \wedge \gamma)}{O(\alpha|\beta \wedge \beta'\backslash\neg\gamma)}$$

$$\frac{O_{\gamma}(\alpha|\beta), \diamondsuit(\alpha \wedge \beta \wedge \gamma \wedge \gamma')}{O_{\gamma \wedge \gamma'}(\alpha|\beta)} \qquad \frac{O(\alpha|\beta\backslash\neg\gamma), \diamondsuit(\alpha \wedge \beta \wedge \gamma \wedge \gamma')}{O(\alpha|\beta\backslash\neg\gamma \vee \neg\gamma')}$$

A side-effect of the definition of the preference-based contextual obligation and restricted strengthening of the preferences is the following weakening of the consequent. It illustrates the interaction between antecedent and context, which is useful in the analysis of contrary-to-duty reasoning. For example, consider the second best plan paradox of Example 2. From the obligation *'give an aspirin today and give one tomorrow'* $O_{\top}(a_1 \wedge a_2 \mid \top)$ we can derive the obligation *'give an aspirin tomorrow unless no-one is given today'* $O(a_2 \mid \top \backslash \neg a_1)$, but not an obligation *'give an aspirin tomorrow'* simpliciter $O_{\top}(a_2|\top)$.

$$\frac{O_{\gamma}(\alpha \wedge \alpha'|\beta)}{O_{\gamma \wedge \alpha'}(\alpha|\beta)} \qquad \frac{O(\alpha \wedge \alpha'|\beta\backslash\neg\gamma)}{O(\alpha|\beta\backslash\neg\gamma \vee \neg\alpha')}$$

Whether we use $O_{\gamma}(\alpha \mid \beta)$ or $O(\alpha \mid \beta\backslash\gamma)$ is arbitrary, of course, but in many cases the distinction between context and condition (antecedent) is unclear. The interpretation of the first type is therefore more difficult, and we advise to use the second type. In the following section we axiomatize a preference-based contextual deontic logic.

3 AXIOMATIZATION

Contextual obligations are formalized in Boutilier's modal preference logic CT4O, a bimodal propositional logic of inaccessible worlds.[4] For the details and completeness proof of this logic see [Boutilier, 1994]. The advantages of our formalization in a modal framework, where an operator is given by a definition in an underlying logic, is that we get an axiomatization for free! We do not have to look for a sound and complete set of inference rules and axiom schemata, because we simply take the axiomatization of the underlying logic together with the new definition. In other words, the problem of finding a sound and complete axiomatization is replaced by the problem of finding a definition of a contextual obligation in terms of a monadic modal preference logic. In the logic we abstract from actions, time and individuals.

DEFINITION 4 (CT4O). The logic CT4O is a propositional bimodal system with the two normal modal connectives $\Box$ and $\overleftarrow{\Box}$. Dual 'possibility' connectives $\Diamond$ and $\overleftarrow{\Diamond}$ are defined as usual and two additional modal connectives $\overleftrightarrow{\Box}$ and $\overleftrightarrow{\Diamond}$ are defined as follows.

$$\Diamond\alpha =_{def} \neg\Box\neg\alpha \qquad \overleftrightarrow{\Box}\alpha =_{def} \Box\alpha\wedge\overleftarrow{\Box}\alpha$$

$$\overleftarrow{\Diamond}\alpha =_{def} \neg\overleftarrow{\Box}\neg\alpha \qquad \overleftrightarrow{\Diamond}\alpha =_{def} \Diamond\alpha\vee\overleftarrow{\Diamond}\alpha$$

CT4O is axiomatized by the following axioms and inference rules.

K	$\Box(\alpha\to\beta)\to(\Box\alpha\to\Box\beta)$	**Nes**	From α infer $\overleftrightarrow{\Box}\alpha$
K′	$\overleftarrow{\Box}(\alpha\to\beta)\to(\overleftarrow{\Box}\alpha\to\overleftarrow{\Box}\beta)$	**MP**	From $\alpha\to\beta$ and α infer β
T	$\Box\alpha\to\alpha$		
4	$\Box\alpha\to\Box\Box\alpha$		
H	$\overleftrightarrow{\Diamond}(\Box\alpha\wedge\overleftarrow{\Box}\beta)\to\overleftrightarrow{\Box}(\alpha\vee\beta)$		

Kripke models $M=\langle W,\leq,V\rangle$ for CT4O consist of W, a set of worlds, $\leq$, a binary transitive and reflexive accessibility relation, and V, a valuation of the propositional atoms in the worlds. The partial pre-ordering $\leq$ expresses preferences: $w_1\leq w_2$ if and only if w_1 is as least as preferable as w_2. The modal connective $\Box$ refers to accessible worlds and the modal connective $\overleftarrow{\Box}$ to inaccessible worlds.

$$M,w\models\Box\alpha \text{ iff } \forall w'\in W \text{ if } w'\leq w, \text{ then } M,w'\models\alpha$$

$$M,w\models\overleftarrow{\Box}\alpha \text{ iff } \forall w'\in W \text{ if } w'\not\leq w, \text{ then } M,w'\models\alpha$$

[4] We can also use standard bimodal logics – of which the models contain two accessibility relations – axiomatized by $\overleftrightarrow{\Box}\alpha\to\Box\alpha$. In that case the $\overleftrightarrow{\Box}$ operator is not defined by a universal relation, but by an equivalence relation. The axiomatization will be simpler but the semantic presentation will be less clear, because we always have to restrict our focus to the equivalence class accessible from the actual world.

Contextual obligations are defined in CT4O as follows. In this paper, we do not discuss the properties of $\succ$ but we focus on the properties of the contextual obligations.[5]

DEFINITION 5 (CDL). The logic CDL is the logic CT4O extended with the following definitions of contextual obligations. The contextual obligation 'α should be the case if β is the case unless γ is the case', written as $O(\alpha|\beta\backslash\gamma)$, is defined as a strong preference of $\alpha \wedge \beta \wedge \neg\gamma$ over $\neg\alpha \wedge \beta$.

$$\begin{array}{lcl} \alpha_1 \succ \alpha_2 & =_{def} & \overleftrightarrow{\Box}(\alpha_1 \to \Box\neg\alpha_2) \wedge \overleftrightarrow{\Diamond}\alpha_1 \\ O(\alpha|\beta\backslash\gamma) & =_{def} & (\alpha \wedge \beta \wedge \neg\gamma) \succ (\neg\alpha \wedge \beta) \\ & = & \overleftrightarrow{\Box}((\alpha \wedge \beta \wedge \neg\gamma) \to \Box(\beta \to \alpha)) \wedge \overleftrightarrow{\Diamond}(\alpha \wedge \beta \wedge \neg\gamma) \\ O_\gamma(\alpha|\beta) & =_{def} & O(\alpha|\beta\backslash\neg\gamma) \end{array}$$

From the definitions follows immediately the following satisfiability conditions for the two additional modal connectives: $M, w \models \overleftrightarrow{\Box}\, \alpha$ if and only if $\forall w' \in W$ $M, w' \models \alpha$ and $M, w \models \overleftrightarrow{\Diamond}\, \alpha$ if and only if $\exists w' \in W$ $M, w' \models \alpha$. As a consequence, the truth value of a contextual obligation does not depend on the world in which the obligation is evaluated. For a model $M = \langle W, \leq, V\rangle$ we have $M \models O(\alpha \mid \beta\backslash\gamma)$ (i.e. for all worlds $w \in W$ we have $M, w \models O(\alpha \mid \beta\backslash\gamma)$) if and only if there is a world $w \in W$ such that $M, w \models O(\alpha|\beta\backslash\gamma)$.

The following proposition shows the truth conditions of contextual obligations.

PROPOSITION 6 (Contextual obligation). *Let $M = \langle W, \leq, V\rangle$ be a CT4O model and let $|\,\alpha\,|$ be the set of worlds that satisfy α. For a world $w \in W$, we have $M, w \models O(\alpha|\beta\backslash\gamma)$ if and only if for all $w_1 \in |\alpha \wedge \beta \wedge \neg\gamma|$ and all $w_2 \in |\neg\alpha \wedge \beta|$ we have $w_2 \not\leq w_1$, and such a world w_1 exists.*

Proof. Follows directly from the definition of $\succ$. ■

The preference-based dyadic obligations $O(\alpha \mid \beta) =_{def} O_\top(\alpha \mid \beta)$ are weak variants of the obligations defined in so-called Prohairetic Deontic Logic (PDL) in [van der Torre & Tan, 1998a]. In PDL, obligations have an additional restriction and are defined by

$$O_{PDL}(\alpha|\beta) =_{def} (\alpha \wedge \beta \succ \neg\alpha \wedge \beta) \wedge I(\alpha|\beta)$$

where $I(\alpha \mid \beta)$ stands for 'the ideal (or preferred) β worlds are α worlds.' As a consequence of the additional condition, dilemmas are inconsistent in PDL. Nevertheless, weak variants of the properties of PDL can be given for CDL. We therefore immediately have that CDL has restricted strengthening of the antecedent,

[5] The preference relation $\succ$ is quite weak. For example, it is not anti-symmetric (we cannot derive $\neg(\alpha_2 \succ \alpha_1)$ from $\alpha_1 \succ \alpha_2$) and it is not transitive (we cannot derive $\alpha_1 \succ \alpha_3$ from $\alpha_1 \succ \alpha_2$ and $\alpha_2 \succ \alpha_3$). The lack of these properties is the result of the fact that we do not have connected orderings. Moreover, this a-connectedness is crucial for our preference-based deontic logics, see [Tan & van der Torre, 1996; van der Torre & Tan, 1998a].

restricted deontic detachment, and a restricted conjunction rule. For the details see [van der Torre & Tan, 1998a; van der Torre & Tan, 1999]. The following proposition shows several properties of contextual obligations relevant for the analysis of contextual reasoning. In contrast to PDL, CDL also has a kind of weakening of the consequent.

PROPOSITION 7 (Theorems of CDL). *The logic CT4O validates the following theorems.*

$$\begin{array}{ll} \textbf{RSA}: & (O(\alpha|\beta_1\backslash\gamma)\wedge \overset{\leftrightarrow}{\Diamond}(\alpha\wedge\beta_1\wedge\beta_2\wedge\neg\gamma)) \rightarrow O(\alpha|\beta_1\wedge\beta_2\backslash\gamma) \\ \textbf{RSC}: & (O(\alpha|\beta\backslash\gamma_1)\wedge \overset{\leftrightarrow}{\Diamond}(\alpha\wedge\beta\wedge\neg\gamma_1\wedge\neg\gamma_2)) \rightarrow O(\alpha|\beta\backslash\gamma_1\vee\gamma_2) \\ \textbf{WC}: & O(\alpha_1\wedge\alpha_2|\beta\backslash\gamma) \rightarrow O(\alpha_1|\beta\backslash\gamma\vee\neg\alpha_2) \end{array}$$

Proof. The theorems can easily be proven in the preferential semantics. Consider **WC**. Assume $M \models O(\alpha_1 \wedge \alpha_2 \mid \beta\backslash\gamma)$. Let $W_1 = \mid \alpha_1 \wedge \alpha_2 \wedge \beta \wedge \neg\gamma \mid$ and $W_2 = \mid\neg(\alpha_1 \wedge \alpha_2) \wedge \beta\mid$, and $w_2 \not\leq w_1$ for $w_1 \in W_1$ and $w_2 \in W_2$. Moreover, let $W_1' = |\alpha_1 \wedge \beta \wedge \neg(\gamma \vee \neg\alpha_2)|$ and $W_2' = |\neg\alpha_1 \wedge \beta|$. We have $w_2 \not\leq w_1$ for $w_1 \in W_1'$ and $w_2 \in W_2'$, because $W_1 = W_1'$ and $W_2' \subseteq W_2$. Thus, $M \models O(\alpha_1|\beta\backslash\gamma \vee \neg\alpha_2)$. Verification of the other theorems is left to the reader. ■

To illustrate the properties of CDL, we compare it with Hansson's minimizing dyadic deontic logic. First we recall some well-known definitions and properties of this logic. In Hansson's classical preference semantics [Hansson, 1971], as studied by Lewis [1974], a dyadic obligation, which we denote by $O_{HL}(\alpha \mid \beta)$, is true in a model if and only if $I(\alpha \mid \beta)$ is true: 'the ideal (or preferred) β worlds satisfy α.' A weaker version of this definition for our models (which are not necessarily connected, i.e. we can have incomparable worlds) is that $O^w_{HL}(\alpha \mid \beta)$ is true in a model if and only if 'there is an *equivalence class* of minimal (or preferred) β worlds that satisfy α.' These obligations are not closed under the conjunction rule $Op \wedge Oq \rightarrow O(p \wedge q)$ and therefore allow for a consistent representation of dilemmas.

DEFINITION 8 (Minimizing obligation). Let $M = \langle W, \leq, V\rangle$ be a Kripke model and $\mid\alpha\mid$ be the set of all worlds of W that satisfy α. The weak Hansson-Lewis obligation 'α should be the case if β is the case', written as $O^w_{HL}(\alpha|\beta)$, is defined as follows.

$$O^w_{HL}(\alpha|\beta) \quad =_{def} \quad \overset{\leftrightarrow}{\Diamond}(\beta \wedge \Box(\beta \rightarrow \alpha))$$

The model M satisfies the weak Hansson-Lewis obligation 'α should be the case if β is the case', written as $M \models O^w_{HL}(\alpha|\beta)$, if and only if there is a world $w_1 \in|\alpha \wedge \beta|$ such that for all $w_2 \in|\neg\alpha \wedge \beta|$ we have $w_2 \not\leq w_1$. The following proposition shows that the expression $O^w_{HL}(\alpha|\beta)$ corresponds to a weak Hansson-Lewis minimizing obligation. For simplicity, we assume that there are no infinite descending chains.

PROPOSITION 9. *Let $M = \langle W, \leq, V\rangle$ be a CT4O model, such that there are no infinite descending chains. As usual, we write $w_1 < w_2$ for $w_1 \leq w_2$ and not $w_2 \leq w_1$, and $w_1 \sim w_2$ for $w_1 \leq w_2$ and $w_2 \leq w_1$. A world w is a minimal β-world, written as $M, w \models_{\leq} \beta$, if and only if $M, w \models \beta$ and for all $w' < w$ holds $M, w' \not\models \beta$. A set of worlds is an equivalence class of minimal β-worlds, written as E_β, if and only if there is a w such that $M, w \models_{\leq} \beta$ and $E_\beta = \{w' \mid M, w' \models \beta \text{ and } w \sim w'\}$. We have $M \models O^{w}_{HL}(\alpha|\beta)$ if and only if there is an E_β such that $E_\beta \subseteq |\alpha|$.*

Proof. $\Leftarrow$ Follows directly from the definitions. Assume there is a w such that $M, w \models_{\leq} \beta$ and $E_\beta = \{w' \mid M, w' \models \beta$ and $w \sim w'\}$ and $E_\beta \subseteq |\alpha|$. For all $w_2 \in |\neg\alpha \wedge \beta|$ we have $w_2 \not\leq w$.

$\Rightarrow$ Assume that there is a world $w_1 \in |\alpha \wedge \beta|$ such that for all $w_2 \in |\neg\alpha \wedge \beta|$ we have $w_2 \not\leq w_1$. Let w be a minimal β-world such that $M, w \models_{\leq} \beta$ and $w \leq w_1$ (that exists because there are no infinite descending chains), and let $E_\beta = \{w' \mid M, w' \models \beta$ and $w \sim w'\}$. ■

Now we are ready to compare our contextual deontic logic with Hansson's dyadic deontic logic. The following proposition shows that under a certain condition, the contextual obligation $O(\alpha|\beta\backslash\gamma)$ is true in a model if and only if a set of the weak Hansson-Lewis minimizing obligations $O^{w}_{HL}(\alpha|\beta')$ is true in the model. The condition is that the formal language is expressive enough to distinguish between each two worlds.

PROPOSITION 10. *Let $M = \langle W, \leq, V\rangle$ be a CT4O model, that has no worlds that satisfy the same propositional sentences. Hence, we identify the set of worlds with a set of propositional interpretations, such that there are no duplicate worlds. We have $M \models O(\alpha|\beta\backslash\gamma)$ if and only if there are $\alpha \wedge \beta \wedge \neg\gamma$ worlds, and for all propositional β' such that $M \models \ddot{\Box}\,(\beta' \rightarrow \beta)$ and $M \not\models \ddot{\Box}\,(\beta' \rightarrow \alpha \rightarrow \gamma)$, we have $M \models O^{w}_{HL}(\alpha|\beta')$.*

Proof. $\Rightarrow$ There exists a β' world that satisfies $\alpha \wedge \beta \wedge \neg\gamma$, because $M \not\models \ddot{\Box}\,(\beta' \rightarrow \alpha \rightarrow \gamma)$. Each $\alpha \wedge \beta \wedge \neg\gamma$ world satisfies $\beta \wedge \Box(\beta \rightarrow \alpha)$, as a consequence of the definition of $O(\alpha|\beta\backslash\gamma)$.

$\Leftarrow$ Every world is characterized by a unique propositional sentence. Let $\overline{w}$ denote the sentence that uniquely characterizes world w. Proof by contraposition. If $M \not\models O(\alpha|\beta\backslash\gamma)$, then either there are no $\alpha \wedge \beta \wedge \neg\gamma$ worlds (in which case the proof is trivial), or there are w_1, w_2 such that $M, w_1 \models \alpha \wedge \beta \wedge \neg\gamma$, $M, w_2 \models \neg\alpha \wedge \beta$ and $w_2 \leq w_1$. Choose $\beta' = \overline{w_1} \vee \overline{w_2}$. The world w_2 is an element of the preferred β' worlds, because there are no duplicate worlds. (If duplicate worlds are allowed, then there could be a β' world w_3 which is a duplicate of w_1, and which is strictly preferred to w_1 and w_2.) We have $M, w_2 \not\models \alpha$ and therefore $M \not\models O^{w}_{HL}(\alpha|\beta')$. ■

The examples in the following section illustrate the CDL analysis of the contrary-to-duty paradoxes.

4 EXAMPLES: THE CONTRARY-TO-DUTY PARADOXES

Preference-based deontic logics have proven useful to understand the nature of contrary-to-duty (CTD) and according-to-duty (ATD) reasoning. An obligation $O(\alpha \mid \beta)$ is a CTD (ATD) obligation of the *primary* obligation $O(\alpha_1 \mid \beta_1)$ if and only if $\beta \wedge \alpha_1$ is inconsistent (β propositionally implies α_1), as represented in Figure 1.

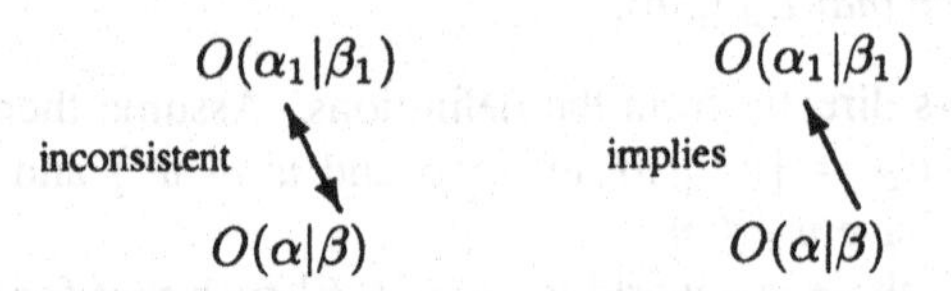

Figure 1. $O(\alpha|\beta)$ is a CTD / ATD obligation of $O(\alpha_1|\beta_1)$

Some typical examples are given below. The first two sets show how the preference-based representation reveals the structure of the CTDs and ATDs. In particular, it shows that they are *not* sets of conflicting preferences. Moreover, the third set TRANS combines the structural properties of CTD and ATD. It plays an important role in defeasible deontic logic, see Section 5.

$CTD:$

$O(a_1|\top), O(a_2|\neg a_1) \qquad a_1 \succ \neg a_1, (a_2 \wedge \neg a_1) \succ (\neg a_2 \wedge \neg a_1)$

$ATD:$

$O(a_1|a_1 \vee a_2), O(a_1 \vee a_2|\top) \qquad a_1 \succ (\neg a_1 \wedge a_2), (a_1 \vee a_2) \succ (\neg a_1 \wedge \neg a_2)$

$TRANS:$

$O(a_1|a_1 \vee a_2), O(a_1 \vee a_2|\neg a_1) \qquad a_1 \succ (\neg a_1 \wedge a_2), (\neg a_1 \wedge a_2) \succ (\neg a_1 \wedge \neg a_2)$

The following two examples reconsider the contrary-to-duty examples discussed in the introduction.

EXAMPLE 11 (Second best plan paradox, continued). Consider the obligations *'Dr Denton ought to give an aspirin today and an aspirin tomorrow'* $O(a_1 \wedge a_2|\top)$ and *'if he does not give an aspirin today, then he ought not to give one tomorrow'* $O(\neg a_2 \mid \neg a_1)$. The context is used to assure that counterintuitive derivations are blocked. If the context of $O(a_1 \wedge a_2|\top)$ is 'always,' then the context of the derived obligation $O(a_2 \mid \top)$ is 'unless $\neg a_1$.' A simple consistency check ensures that $O(a_2|\top)$ 'unless $\neg a_1$' cannot be used to derive $O(a_2|\neg a_1)$, because the antecedent of the latter violates the context of the former. We can derive $O(a_2|\top \setminus \neg a_1)$, but we cannot derive $O_\gamma(a_2|\neg a_1)$ for any γ. The unless clause $\neg a_1$ prevents that the obligation from the ideal best plan context is strengthened to the second best plan context.

EXAMPLE 12 (Cottage housing regulations, continued). Consider the obligations *'there ought to be no fence'* $O(\neg f \mid \top)$, *'if there is a fence, then it ought to be white'* $O(w \wedge f|f)$, *'there ought not to be a dog'* $O(\neg d|\top)$, *'if there is no dog, then there ought to be no sign'* $O(\neg s \mid \neg d)$ and *'if there is a dog, then there ought to be a sign'* $O(s|d)$. We cannot derive $O(\neg(f \wedge w)|f)$ from $O(\neg f|\top)$, as desired. Moreover, we can derive $O(\neg s \wedge \neg d|\top)$, but we cannot derive $O(\neg s|d)$ from $O(\neg d|\top)$ and $O(\neg s|\neg d)$, as desired. Finally, we can derive the obligation *'to have no sign'* in the ideal context there is no dog $O_{\neg d}(\neg s|\top)$. In other words, *'there should not be a sign, unless there is a dog'* $O(\neg s|\top \backslash d)$. Note that the latter two obligations are much stronger than the premise $O(\neg s|\neg d)$. On the other hand, we cannot derive an obligation *'there should not be a sign'* simpliciter $O_\top(\neg s|\top)$.

The cottage housing regulations in Example 12 illustrate that in CDL $O(\alpha_1 \mid \neg\alpha_2)$ cannot be derived from $O(\alpha_1 \wedge \alpha_2 \mid \top)$, because we cannot derive a CTD obligation from its primary obligation. We can derive $O_{\alpha_2}(\alpha_1 \mid \top)$ from $O_\top(\alpha_1 \wedge \alpha_2)$, but this derived obligation cannot be used to derive the counterintuitive $O(\alpha_1 \mid \neg\alpha_2)$ due to the context. The following example illustrates that the derivation of the obligation $O(\alpha_1 \mid \neg\alpha_2)$ from the obligation $O(\alpha_1 \wedge \alpha_2 \mid \top)$ is a fundamental problem underlying several contrary-to-duty paradoxes. Hence, the underlying problem of the contrary-to-duty paradoxes is that a contrary-to-duty obligation can be derived from its primary obligation.

EXAMPLE 13 (Contrary-to-duty paradoxes). Assume a dyadic deontic logic that validates at least substitution of logical equivalents and the following inference patterns *Restricted Strengthening of the Antecedent* (RSA), *Weakening of the Consequent* (WC), *Conjunction of the Consequent* (AND) and a version of *Deontic Detachment* (DD), in which $\overset{\leftrightarrow}{\Diamond}$ is a modal operator and $\overset{\leftrightarrow}{\Diamond}\phi$ is true for all consistent propositional formulas ϕ.

$$\text{RSA}: \frac{O(\alpha|\beta_1), \overset{\leftrightarrow}{\Diamond}(\alpha \wedge \beta_1 \wedge \beta_2)}{O(\alpha|\beta_1 \wedge \beta_2)} \qquad \text{WC}: \frac{O(\alpha_1|\beta)}{O(\alpha_1 \vee \alpha_2|\beta)}$$

$$\text{AND}: \frac{O(\alpha_1|\beta), O(\alpha_2|\beta)}{O(\alpha_1 \wedge \alpha_2|\beta)} \qquad \text{DD}: \frac{O(\alpha|\beta), O(\beta|\gamma)}{O(\alpha \wedge \beta|\gamma)}$$

Furthermore, consider the sets

$$S = \{O(\neg k|\top), O(g|k), \vdash g \to k, k\}$$

$$S' = \{O(a|\top), O(t|a), O(\neg t|\neg a), \neg a\}$$

$$S'' = \{O(\neg a|\top), O(a \vee p|\top), O(\neg p|a)\}$$

$$S''' = \{O(\neg r \wedge \neg g|\top), O(r|g)\}$$

S formalizes Forrester's paradox [Forrester, 1984] when k is read as 'Smith kills Jones' and g as 'he kills him gently,' S' formalizes Chisholm's paradox [Chisholm, 1963] when a is read as 'a certain man going to the assistance of his neighbors' and t as 'the man telling his neighbors that he will come,' S'' formalizes the apples-and-pears example [Tan & van der Torre, 1996] when a is read as 'buying apples' and p as 'buying pears,' and finally, S''' formalizes a part of the Reykjavik scenario [Belzer, 1986; van der Torre, 1994] when r is read as telling a secret to Reagan and g as telling it to Gorbatsjov. The last obligation of each premise set is a contrary-to-duty obligation of the first obligation of the set, because its antecedent is contradictory with the consequent of the latter. The paradoxical consequences of the sets of obligations are represented in Figure 2. The underlying problem of the counterintuitive derivations is the derivation of the obligation $O(\alpha_1|\neg\alpha_2)$ from $O(\alpha_1 \wedge \alpha_2|\top)$ by WC and RSA: respectively the derivation of $O(\neg(g \wedge k)|k)$ from $O(\neg k|\top)$, $O(t|\neg a)$ from $O(a \wedge t|\top)$, $O(p|a)$ from $O(\neg a \wedge p|\top)$, and $O(\neg r|g)$ from $O(\neg r \wedge \neg g|\top)$. These derivations are blocked in CDL.

$$\dfrac{\dfrac{\dfrac{O(\neg k|\top)}{O(\neg g|\top)}\;\text{WC}}{O(\neg g|k)}\;\text{RSA} \qquad O(g|k)}{O(\neg g \wedge g|k)}\;\text{AND}$$

$$\dfrac{\dfrac{\dfrac{\dfrac{O(t|a) \quad O(a|\top)}{O(a \wedge t|\top)}\;\text{DD}}{O(t|\top)}\;\text{WC}}{O(t|\neg a)}\;\text{RSA} \qquad O(\neg t|\neg a)}{O(t \wedge \neg t|\neg a)}\;\text{AND}$$

$$\dfrac{\dfrac{\dfrac{\dfrac{O(\neg a|\top) \quad O(a \vee p|\top)}{O(\neg a \wedge p|\top)}\;\text{AND}}{O(p|\top)}\;\text{WC}}{O(p|a)}\;\text{RSA} \qquad O(\neg p|a)}{O(p \wedge \neg p|a)}\;\text{AND}$$

$$\dfrac{\dfrac{\dfrac{O(\neg r \wedge \neg g|\top)}{O(\neg r|\top)}\;\text{WC}}{O(\neg r|g)}\;\text{RSA} \qquad O(r|g)}{O(r \wedge \neg r|g)}\;\text{AND}$$

Figure 2. Four contrary-to-duty paradoxes

5 DEFEASIBLE DEONTIC LOGIC

A defeasible deontic logic is a deontic logic in which the obligations do not have unrestricted strengthening of the antecedent.[6] Defeasibility, i.e. lack of strengthening of the antecedent, has been used to analyze the following four phenomena. The role of defeasibility in deontic reasoning is complex, because different types of defeasibility have to be distinguished for the different phenomena, see [van der Torre & Tan, 1995; van der Torre & Tan, 1997].

Prima facie obligations: overridden defeasibility. The first and most obvious phenomena is prima facie reasoning, or, more generally, the reasoning about obligations that can be overridden by stronger obligations. For example, *'you should keep your promises,'* but *'one may break a promise in order to prevent a disaster'* [Ross, 1930], *'normally you should not eat with your fingers,'* but *'if you are served asparagus, then normally you should eat with your fingers'* [Horty, 1994], *'normally there should not be a fence,'* but *'if the house in next to a cliff then normally you are permitted to have a fence'* [Prakken & Sergot, 1996], etc.

Dilemmas: conflict defeasibility. Horty [1994; 1997] gives a normative interpretation of Reiter's default logic, which is used to analyze dilemmas. He shows formal links between Reiter's default logic and Van Fraassen's analysis of dilemmas [van Fraassen, 1973].

Unresolved conflicts: conflict defeasibility. Related to dilemmas is the issue of unresolved conflicts. For example, the two obligations of Von Wright's window problem [von Wright, 1963], *'the window ought to be closed if it rains'* $O(c\,|\,r)$ and *'it should be open if the sun shines'* $O(\neg c\,|\,s)$, conflict when it rains and the sun shines. We cannot derive $O(c|r \wedge s)$ nor $O(\neg c|r \wedge s)$. The simplest example consists of two unrelated obligations *'be polite'* Op and *'be honest'* Oh, that conflict if we only consider $p \leftrightarrow \neg h$ worlds. Consequently, in some (preference-based) logics we cannot derive $O(p|p \leftrightarrow \neg h)$.

Violability: factual defeasibility. In this paper we illustrated how factual defeasibility can be used to analyze CTD and ATD reasoning. CTD and ATD reasoning is context-dependent, and contextual reasoning is defeasible.

In this paper we analyzed the relation between contextual reasoning and deontic reasoning in which obligations cannot be overridden by other obligations. The relation between contextual reasoning and *defeasible* deontic reasoning in which obligations can be overridden is much more complex. It can be analyzed

[6] To analyze the relation between obligations and defeasibility, we can either look at non-monotonicity in a monadic deontic logic, or at lack of strengthening of the antecedent in (an extension of) dyadic deontic logic. In this paper we followed the latter approach, based on Alchourrón's definition of a defeasible conditional [Alchourrón, 1993]. A defeasible conditional is a conditional that does not have strengthening of the antecedent.

by making strengths and impacts of obligations explicit in the formal language. A well-known example of the context-dependence of obligations is related to the specificity principle. In most cases the strengths of obligations are unrestricted, but if an obligation is more specific than and conflicting with another obligation, then the strength of the former is higher than the strength of the latter. In general, strengths of obligations are restricted to resolve conflicts. Moreover, there are types of context-dependence that do not depend on the strengths of obligations or conflicts between obligations, as illustrated in the example below. In the framework of [van der Torre & Weydert, 1998] we therefore distinguish between the strengths of obligations and what we call their impact. The following example illustrates the two different types of context-dependence and the distinction between strength and impact.

EXAMPLE 14. Consider the two sets $S_1 = \{O(\neg s \mid \top), O(s \mid i)\}$ and $S_2 = \{O(p \mid p \vee c), O(c \mid c \vee h)\}$, with background knowledge $\top \leftrightarrow (p \vee c \vee h)$, $\neg(p \wedge c)$, $\neg(p \wedge h)$, $\neg(c \wedge h)$ (the three variables p, c and h are mutually exclusive and exhaustive), and with for each set strengths s_1 and s_2, and impacts α_1 and α_2. The first type of context-dependence (set S_1) is a consequence of resolving conflicts and the second type of context-dependence (set S_2) follows from the relation between obligations and preferences; it is *not* a consequence of resolving conflicts. As represented in Figure 3, the representation of obligations suggests that S_2 is a conflict, because the consequents of the obligations are contradictory. However, this is not a sufficient reason for two obligations to conflict. In contrast, the set S_2 is conflict free, because the antecedents of both obligations are only true if c is the case, and in that case the first obligation can no longer be fulfilled. This is especially clear when the obligations are rewritten as preferences. An obligation $O(a|b)$ is formalized by a preference of the fulfilled obligation $a \wedge b$ over the violated obligation $\neg a \wedge b$, which can be written as $O(a|b) =_{def} a \wedge b \succ \neg a \wedge b$. Hence, the set S_2 can also be written as $\{p \succ c, c \succ h\}$. The representation of preferences reveals that S_2 is conflict-free, because $p \wedge h$ is inconsistent. There are not two states such that the first is preferred to the second one by the first preference, and vice versa by the second preference. The reason that the impact of $O(c|c \vee h)$ is higher than the impact of $O(p|p \vee c)$ also follows from the formalization of the obligations as preferences. The context-dependent meaning of the obligations is such that it reflects the preferences.

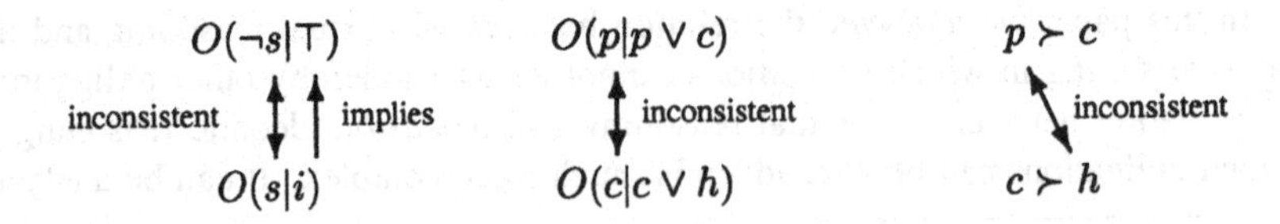

Figure 3. Context-dependent obligations: specificity and conflict-free

6 RELATED RESEARCH

There are two types of dyadic deontic logics, dependent on how the antecedent is interpreted. The first type, as advocated by Chellas [Chellas, 1974; Alchourrón, 1993], defines a dyadic obligation in terms of a monadic obligation by $O(\alpha|\beta) =_{def} \beta > O\alpha$, where '>' is a strict implication. These dyadic deontic logics have strengthening of the antecedent, but they cannot represent the contrary-to-duty paradoxes in a consistent way. Dyadic deontic logics of the second type, as introduced by Hansson [1971] and further investigated by Lewis [1974], do not have strengthening of the antecedent and they therefore can represent the paradoxes in a consistent way.

In [Tan & van der Torre, 1996] we proposed the two-phase deontic logic 2DL to combine restricted strengthening of the antecedent (RSA) and weakening of the consequent (WC). In 2DL, the first phase validates RSA but not WC, and the second phase validates WC but not RSA. 2DL can be used to analyze the contrary-to-duty paradoxes, because the obligation $O(\alpha_1 \mid \neg\alpha_2)$ cannot be derived from $O(\alpha_1 \wedge \alpha_2 \mid \top)$ by respectively WC and RSA. Proof-theoretically, such sequencing of derivations is rather complicated, but semantically, the two-phase approach simply means that first an ordering has to be constructed before it can be used for minimization.

Prakken and Sergot formalize the context as a propositional sentence, and write contextual obligations 'α ought to be the case in context β' as $O_\beta\alpha$. In a second paper, they extend the classical dyadic deontic logics of Hansson [1971] and Lewis [1974]. They look for a consistency check to block strengthening of the context [Prakken & Sergot, 1996] or strengthening of the antecedent [Prakken & Sergot, 1997]. Moreover, in [Prakken & Sergot, 1997] they show it is impossible to construct such a consistency check in the classical Hansson-Lewis semantics.

Makinson [1999] proposes a labeled deontic logic in which the inference relation is relative to a set of explicit promulgations following [Alchourrón & Bulygin, 1981]. He uses an hypothesis that ensures that obligations are not strengthened to another context. However, this hypothesis is not represented in the formal language.

The contextual obligations $O_\gamma(\alpha|\beta)$ as defined in this paper are closely related to labeled obligations $O(\alpha|\beta)_L$, to be read as 'α ought to be (done) if β is (done) against the background of L' [van der Torre & Tan, 1995; 1997; van der Torre, 1998a; 1998b]. Labeled deontic logic LDL is a labeled deductive system as introduced by Gabbay in [Gabbay, 1996]. It is a powerful technique to define complex inductive definitions with several arguments. The label formalizes the context in which an obligation is derived. The advantage of CDL over LDL is that it has a semantics, and that its language contains negations and disjunctions of obligations, as well as facts (and thus violations).

7 CONCLUSIONS

Contextual deontic logic CDL is a defeasible deontic logic that has factual defeasibility but not overridden defeasibility. This limited expressive power is useful to analyze contrary-to-duty reasoning without confusing factual defeasibility with overridden defeasibility, i.e. without confusing violations with exceptions. We also showed how the notorious contrary-to-duty paradoxes of deontic logic can be formalized in CDL.

There are two main drawbacks of the logic. First, it satisfies the preference-based theorems such as $O(\alpha|\alpha)$. Second, it does not support reasoning by cases, because the disjunction rule for the antecedent is not valid. The development of modifications and extensions of the logic that do not have these drawbacks is subject of present research.

Leendert W. N. van der Torre
Max Planck Institute for Computer Science, Germany.

Yao-Hua Tan
Erasmus University Rotterdam, The Netherlands.

REFERENCES

[Alchourrón & Bulygin, 1981] Alchourrón, C., and Bulygin. 1981. The expressive conception of norms. In Hilpinen, R., ed., *New Studies in Deontic Logic: Norms, Actions and the Foundations of Ethics*. D. Reidel. 95–124.

[Alchourrón, 1993] Alchourrón, C. 1993. Philosophical foundations of deontic logic and the logic of defeasible conditionals. In Meyer, J.-J., and Wieringa, R., eds., *Deontic Logic in Computer Science: Normative System Specification*. John Wiley & Sons. 43–84.

[Belzer, 1986] Belzer, M. 1986. A logic of deliberation. In *Proceedings of the Fifth National Conference on Artificial Intelligence (AAAI'86)*, 38–43.

[Boutilier, 1994] Boutilier, C. 1994. Conditional logics of normality: a modal approach. *Artificial Intelligence* 68:87–154.

[Castañeda, 1981] Castañeda, H. 1981. The paradoxes of deontic logic: the simplest solution to all of them in one fell swoop. In Hilpinen, R., ed., *New Studies in Deontic Logic: Norms, Actions and the Foundations of Ethics*. D.Reidel Publishing company. 37–85.

[Chellas, 1974] Chellas, B. 1974. Conditional obligation. In Stunland, S., ed., *Logical Theory and Semantical Analysis*. Dordrecht, Holland: D. Reidel Publishing Company. 23–33.

[Chisholm, 1963] Chisholm, R. 1963. Contrary-to-duty imperatives and deontic logic. *Analysis* 24:33–36.

[Feldmann, 1990] Feldmann, F. 1990. A simpler solution to the paradoxes of deontic logic. In Tomberlin, J., ed., *Philosophical perspectives 4: Action theory and Philosophy of Mind*. Atascadero: Ridgview.

[Forrester, 1984] Forrester, J. 1984. Gentle murder, or the adverbial Samaritan. *Journal of Philosophy* 81:193–197.

[Gabbay, 1996] Gabbay, D. 1996. *Labelled Deductive Systems*, volume 1. Oxford University Press.

[Hansson, 1971] Hansson, B. 1971. An analysis of some deontic logics. In Hilpinen, R., ed., *Deontic Logic: Introductory and Systematic Readings*. Dordrecht, Holland: D. Reidel Publishing Company. 121–147.

[Horty, 1994] Horty, J. 1994. Moral dilemmas and nonmonotonic logic. *Journal of Philosophical Logic* 23:35–65.

[Horty, 1997] Horty, J. 1997. Nonmonotonic foundations for deontic logic. In Nute, D., ed., *Defeasible Deontic Logic*. Kluwer. 17–44.

[Lewis, 1974] Lewis, D. 1974. Semantic analysis for dyadic deontic logic. In Stunland, S., ed., *Logical Theory and Semantical Analysis*. Dordrecht, Holland: D. Reidel Publishing Company. 1–14.

[Loewer & Belzer, 1983] Loewer, B., and Belzer, M. 1983. Dyadic deontic detachment. *Synthese* 54:295–318.

[Makinson, 1999] Makinson, D. 1999. On a fundamental problem of deontic logic. In McNamara, P., and Prakken, H., eds., *Norms, Logics and Information Systems. New Studies on Deontic Logic and Computer Science*. IOS Press.

[McCarty, 1994] McCarty, L. 1994. Defeasible deontic reasoning. *Fundamenta Informaticae* 21:125–148.

[Prakken & Sergot, 1996] Prakken, H., and Sergot, M. 1996. Contrary-to-duty obligations. *Studia Logica* 57:91–115.

[Prakken & Sergot, 1997] Prakken, H., and Sergot, M. 1997. Dyadic deontic logic and contrary-to-duty obligations. In Nute, D., ed., *Defeasible Deontic Logic*. Kluwer. 223–262.

[Reiter, 1980] Reiter, R. 1980. A logic for default reasoning. *Artificial Intelligence* 13:81–132.

[Ross, 1930] Ross, D. 1930. *The Right and the Good*. Oxford University Press.

[Ryu & Lee, 1993] Ryu, Y., and Lee, R. 1993. Defeasible deontic reasoning: A logic programming model. In Meyer, J.-J., and Wieringa, R., eds., *Deontic Logic in Computer Science: Normative System Specification*. John Wiley & Sons. 225–241.

[Tan & van der Torre, 1996] Tan, Y.-H., and van der Torre, L. 1996. How to combine ordering and minimizing in a deontic logic based on preferences. In *Deontic Logic, Agency and Normative Systems. Proceedings of the ΔEON'96. Workshops in Computing*, 216–232. Springer Verlag.

[Thomason, 1981] Thomason, R. 1981. Deontic logic as founded on tense logic. In Hilpinen, R., ed., *New Studies in Deontic Logic: Norms, Actions and the Foundations of Ethics*. D. Reidel. 165–176.

[Tomberlin, 1981] Tomberlin, J. 1981. Contrary-to-duty imperatives and conditional obligation. *Noûs* 16:357–375.

[van der Torre & Tan, 1995] van der Torre, L., and Tan, Y. 1995. Cancelling and overshadowing: two types of defeasibility in defeasible deontic logic. In *Proceedings of the 14th International Joint Conference on Artificial Intelligence (IJCAI'95)*, 1525–1532. Morgan Kaufman.

[van der Torre & Tan, 1997] van der Torre, L., and Tan, Y. 1997. The many faces of defeasibility in defeasible deontic logic. In Nute, D., ed., *Defeasible Deontic Logic*. Kluwer. 79–121.

[van der Torre & Tan, 1998a] van der Torre, L., and Tan, Y. 1998a. Prohairetic Deontic Logic (PDL). In *Logics in Artificial Intelligence*, LNAI 1486, 77–91. Springer.

[van der Torre & Tan, 1998b] van der Torre, L., and Tan, Y. 1998b. The temporal analysis of Chisholm's paradox. In *Proceedings of Fifteenth National Conference on Artificial Intelligence (AAAI'98)*, 650–655.

[van der Torre & Tan, 1998c] van der Torre, L., and Tan, Y. 1998c. An update semantics for prima facie obligations. In *Proceedings of the Thirteenth European Conference on Artificial Intelligence (ECAI'98)*, 38–42.

[van der Torre & Tan, 1999] van der Torre, L., and Tan, Y. 1999. An update semantics for deontic reasoning. In McNamara, P., and Prakken, H., eds., *Norms, Logics and Information Systems. New Studies on Deontic Logic and Computer Science*. IOS Press.

[van der Torre & Weydert, 1998] van der Torre, L., and Weydert, E. 1998. Goals, desires, utilities and preferences. In *Proceedings of the ECAI'98 Workshop Decision Theory meets Artificial Intelligence*.

[van der Torre, 1994] van der Torre, L. 1994. Violated obligations in a defeasible deontic logic. In *Proceedings of the Eleventh European Conference on Artificial Intelligence (ECAI'94)*, 371–375. John Wiley & Sons.

[van der Torre, 1997] van der Torre, L. 1997. *Reasoning about Obligations: Defeasibility in Preference-based Deontic Logic*. Ph.D. Dissertation, Erasmus University Rotterdam.

[van der Torre, 1998a] van der Torre, L. 1998a. Labeled logics of conditional goals. In *Proceedings of the Thirteenth European Conference on Artificial Intelligence (ECAI'98)*, 368–369.

[van der Torre, 1998b] van der Torre, L. 1998b. Phased labeled logics of conditional goals. In *Logics in Artificial Intelligence*, LNAI 1486, 92–106. Springer.

[van Eck, 1982] van Eck, J. 1982. A system of temporally relative modal and deontic predicate logic and its philosophical application. *Logique et Analyse* 100:249–381.

[van Fraassen, 1973] van Fraassen, B. 1973. Values and the heart command. *Journal of Philosophy* 70:5–19.
[von Wright, 1963] von Wright, G. 1963. *The logic of preference*. Edinburgh University Press.
[Yu, 1995] Yu, X. 1995. *Deontic Logic with Defeasible Detachment*. Ph.D. Dissertation, University of Georgia.

FAUSTO GIUNCHIGLIA AND CHIARA GHIDINI

A LOCAL MODELS SEMANTICS FOR PROPOSITIONAL ATTITUDES

1 INTRODUCTION

We are interested in the representation and mechanization of propositional attitudes, and belief in particular, inside complex reasoning programs, e.g. knowledge representation systems, natural language understanding systems, or multiagent systems. Modal logics are the obvious formalism for this goal, the one which has been most widely proposed and studied in the logic and philosophical literature. However, as already discussed in detail in [Giunchiglia and Serafini, 1994; Giunchiglia, 1995], modal logics are hardly used in the existing implemented systems. From the point of view of implementors, the problem is that it is very hard to codify in modal logics all the needed information (e.g. the agents' knowledge, their — usually very different — reasoning capabilities, the — usually very complicated — interactions among them), in a way to have efficient, easy to develop and to maintain implementations. Modal logics allow for compact representations, where one has basically to provide the "appropriate" axioms. This is very elegant, and beautiful from the point of view of the logician. The drawback is that implementors do not find in the formalism enough structure for directly representing all they would like to represent.

Motivated by these considerations, in [Giunchiglia and Serafini, 1994] we have defined various formal systems, called *Hierarchical Multilanguage Belief (HMB) systems*. The idea underlying HMB systems is that a modal logic can be defined in terms of a hierarchy of distinct (that is, not amalgamated) metatheories (metalanguages). In [Giunchiglia and Serafini, 1994] we have produced equivalence results between HMB systems and the most common normal modal logics. In [Giunchiglia *et al.*, 1993; Giunchiglia and Giunchiglia, 1996] we have produced analogous results for the most common non normal modal logics. HMB systems can be basically seen as alternative formulations of modal logics. However, as discussed in detail in [Giunchiglia and Giunchiglia, 1996] (but see also [Giunchiglia and Serafini, 1994; Giunchiglia, 1995]) and briefly hinted below, these systems capture and formalize the current practice in the implementation of complex systems, and behave quite differently from the previous formulations of modal logics.

The goal of this paper is to provide a new semantics, called *Local Models Semantics* [Giunchiglia and Ghidini, 1998], which captures the intuitions which have motivated the definition of HMB systems, and which can be

P. Bonzon, M. Cavalcanti and R. Nossum (eds.), Formal Aspects of Context, 161–174.

used for providing a foundation to the mechanization of belief, as it is implemented in most of the existing systems. Local models semantics develops and generalizes the ideas discussed in [Giunchiglia, 1993]. The idea underlying Local Models Semantics is that each (meta)theory defines a set of first order models; called "local models"; that, e.g., belief is a unary predicate; and that the extension of the belief predicate is computed by enforcing constraints among sets of local models. The motivations for this work are representational. However, the proposed semantics has also some nice technical properties. In particular, in this paper, we show that this semantics scales up very naturally from non-normal to normal modal logics (in particular, modal $\mathcal{K}$), and that it allows for a straightforward treatment of bounded beliefs. Bounded beliefs are almost always needed, or at least very useful, in the mechanization of reasoning about beliefs.

This paper is structured as follows. In Section 2 we briefly present and motivate HMB systems. In Section 3 we give and motivate Local Model Semantics. Finally, in Section 4 we hint soundness and completeness theorems and their proofs.

2 HIERARCHICAL MULTILANGUAGE BELIEF SYSTEMS

Let us consider the situation of a single agent a (usually thought of as the computer itself or as an external observer) who is acting in a world, who has both beliefs about this world and beliefs about its own beliefs and it is able to reason about them (the extension to multiple agents is left as an exercise). We represent beliefs about beliefs by exploiting the notion of *view*. By a view we formalize the mental images that a has of itself, i.e., the set of beliefs that a ascribes to itself.

Views are organized in a chain (see Figure 1). We call a the root view, representing the point of view of the agent a; we let the view aa formalize the beliefs that the agent a ascribes to itself. Iterating the nesting, the view aaa formalizes the view of the agent a about beliefs about its own beliefs, and so on. Notationally we use the natural numbers 0, 1, 2, ... to refer the sequence a, aa, aaa, ... of views. ω refers to the set of natural numbers.

Let us consider only a and aa in Figure 1, that is the agent a and the view that a has of its own beliefs. The idea underlying our formalization is straightforward. The beliefs of a are formalized by a first-order theory. The language of a contains a distinguished predicate B such that $B(\text{"}\phi\text{"})$ holds if and only if a believes that it believes ϕ. The view aa is again formalized as a first-order theory. To obtain the desired behavior, that is to make a able to reason about its own beliefs, it is sufficient to "link" a deduction in the theory of a's beliefs and a deduction in the mental image that a has of itself. Depending on the kind of link a will be a correct believer, that is, it will believe $B(\text{"}\phi\text{"})$ only if it will believe ϕ in the view that it has of itself;

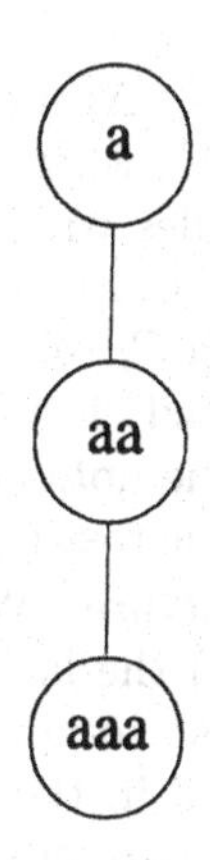

Figure 1. The chain of views

it will be a complete believer, that is, it will believe $B(\text{"}\phi\text{"})$ any time it will believe ϕ in the view that it has of itself; it will be both complete and correct, or neither. Beliefs of arbitrary nesting can be deduced by recursively considering mental images of mental images, up to the necessary depth.

As from the above intuitive description we need to have: (i) multiple first order theories, and (ii) some way to link deduction inside each of them. A Multilanguage System (ML system) [Giunchiglia and Serafini, 1994] is a triple $\langle \{L_i\}_{i\in I}, \{\Omega_i\}_{i\in I}, \Delta \rangle$ where $\{L_i\}_{i\in I}$ is a family of languages, $\{\Omega_i\}_{i\in I}$ is a family of sets of axioms, and Δ is the deductive machinery. Δ contains two kinds of inference rules: ***internal rules*** and ***bridge rules***. Internal rules are inference rules with premises and conclusions in the same language, while bridge rules are inference rules with premises and conclusions belonging to different languages. Notationally, we write inference rules, with, e.g. a single premise, as follows:

$$\frac{i:\phi}{i:\psi}\,ir \qquad\qquad \frac{i:\phi}{j:\psi}\,br$$

where we write $i:\phi$ to mean ϕ and that ϕ is a formula belonging to L_i. We also say that ϕ is an L_i-formula. *ir* is an internal rule while *br* is a bridge rule. We follow Prawitz as in [1965] in the notation and terminology. Derivability in an ML system MS, in symbols $\vdash_{\mathrm{MS}}$ is defined in [Giunchiglia and Serafini, 1994]; roughly speaking it is a generalization of the notion of deduction in Natural Deduction (ND) as given in [Prawitz, 1965]. The generalization amounts to using formulae tagged with the language they belong to.

An ML system $\langle\{L_i\}_{i\in I}, \{\Omega_i\}_{i\in I}, \Delta\rangle$ univocally defines (what we call) a Multicontext system $\langle\{C_i\}_{i\in I}, \Delta_{br}\rangle$, where: $\Delta = \cup_{i\in I}\Delta_i \cup \Delta_{br}$ with Δ_i the set of all and only the internal rules with premises and conclusions in L_i, and Δ_{br} the set of all and only the bridge rules in Δ; $C_i = \langle L_i, \Omega_i, \Delta_i\rangle$ is an axiomatic formal system. Each C_i (also called *context*, to distinguish it from the "usual" notion of "isolated" formal system) formalizes the beliefs of a or of one of its views, while the notion of deduction given above allows us to build deductions inside and across theories.[1]

ML systems have a lot of structure. We can use this structure to impose locally to each view (theory) the language, the basic knowledge, and the deductive capabilities, and also to impose how information propagates among views. As widely discussed in the papers cited above, this gives a lot of modularity. In this paper we are interested in defining systems which are provably equivalent to the "standard" formulations of modal logics. In order to achieve this, we focus our attention to an agent a "on top" of an infinite chain of views. The infinite chain reflects the fact that using modal logics, people models a as an ideal agent able to express and reason about belief formulae with arbitrary nested belief operator, and having arbitrary deep views of its own beliefs. As each view is "on top" an infinite chain and each level corresponds to a level of nesting of the belief predicate B, all the languages used to describe what is true in the views must have the same expressibility. I.e., they are the same language $L(B)$ containing a set P of propositional letters used to express statements about the world, and formulae $B(\text{"}\phi\text{"}), B(\text{"}B(\text{"}\phi\text{"})\text{"}), B(\text{"}B(\text{"}B(\text{"}\phi\text{"})\text{"})\text{"}), \ldots$ used to express beliefs about beliefs.

Formally, let L be a propositional language containing a set P of propositional letters, the symbol for falsity $\bot$, and closed under implication[2]. Then for any natural number $i \in \omega$, L_i is defined as:

- if $\phi \in L$, then $\phi \in L_i$;
- $\bot \in L_i$;
- if $\phi \in L_i$ and $\psi \in L_i$ then $\phi \supset \psi \in L_i$;
- if $\phi \in L_i$ then $B(\text{"}\phi\text{"}) \in L_{i+i}$;
- nothing else is in L_i

[1] ML systems can be thought of as particular Labelled Deductive Systems (LDS's) [Gabbay, 1990; Gabbay, 1994]. In particular ML systems are LDSs where labels are used only to keep track of the language formulae belong to, and where inference rules can be applied only to formulae belonging to the "appropriate" language.

[2] In this paper we use the standard abbreviations from propositional logic, such as $\neg\phi$ for $\phi \supset \bot$, $\phi \vee \psi$ for $\neg\phi \supset \psi$, $\phi \wedge \psi$ for $\neg(\neg\phi \vee \neg\psi)$, $\top$ for $\bot \supset \bot$.

$L(B)$ is defined as follows:

$$L(B) = \bigcup_{n \in \omega} L_n$$

This leads to the following definition of *Hierarchical Multilanguage Belief (HMB) systems.*

DEFINITION 1 (HMB system). Let L be a propositional language. A ML system $\langle \{L_i\}_{i \in I}, \{\Omega_i\}_{i \in I}, \Delta \rangle$ such that for every $i \in \omega$, $L_i = L(B)$, $\Omega_i = \emptyset$, Δ contains the following internal rules:

$$\frac{\begin{matrix}[i : \phi] \\ \vdots \\ i : \psi\end{matrix}}{i : \phi \supset \psi} \supset I_i \qquad \frac{i : \phi \quad i : \phi \supset \psi}{i : \psi} \supset E_i \qquad \frac{\begin{matrix}[i : \neg\phi] \\ \vdots \\ i : \bot\end{matrix}}{i : \phi} \bot_i$$

is called

(i) $\mathcal{R}upr$ if for every $i \in \omega$ the only bridge rule in Δ is $\mathcal{R}upr_i$;
(ii) $\mathcal{R}dwr$ if for every $i \in \omega$ the only bridge rule in Δ is $\mathcal{R}dwr_i$;
(iii) $\mathcal{R}up$ if for every $i \in \omega$ the only bridge rule in Δ is $\mathcal{R}up_i$;
(iv) $\mathcal{R}dw$ if for every $i \in \omega$ the only bridge rule in Δ is $\mathcal{R}dw_i$;

where $\mathcal{R}upr_i$, $\mathcal{R}dwr_i$, $\mathcal{R}up_i$, and $\mathcal{R}dw_i$ are

$$\frac{i+1 : \phi}{i : B(\text{“}\phi\text{”})} \mathcal{R}upr_i \qquad \frac{i : B(\text{“}\phi\text{”})}{i+1 : \phi} \mathcal{R}dwr_i$$

$$\frac{i+1 : \phi}{i : B(\text{“}\phi\text{”})} \mathcal{R}up_i \qquad \frac{i : B(\text{“}\phi\text{”})}{i+1 : \phi} \mathcal{R}dw_i$$

RESTRICTIONS: $\mathcal{R}upr_i$ is applicable if and only if $i+1 : \phi$ does not depend on any assumption $j : \psi$ with index $j \geq i+1$, and $\mathcal{R}dwr_i$ is applicable if and only if $i : B(\text{“}\phi\text{”})$ does not depend on any assumption $j : \psi$ with index $j \leq i$.

For any subset HMB of $\{\mathcal{R}up, \mathcal{R}upr, \mathcal{R}dw, \mathcal{R}dwr\}$, a ML system $\langle \{L_i\}_{i \in I}, \{\Omega_i\}_{i \in I}, \Delta \rangle$ is an HMB system if it is a ζ system for any $\zeta \in$ HMB.

Figure 2 gives a graphical representation of the structure of an HMB system. The theory tagged with 0 is the theory of a, the theory tagged with 1 is the view that a has of its own beliefs, and so on.

From now on, we identify an HMB system with its own set of bridge rules. For instance, we write $\mathcal{R}up$ to mean the HMB system containing only the bridge rule $\mathcal{R}up$.

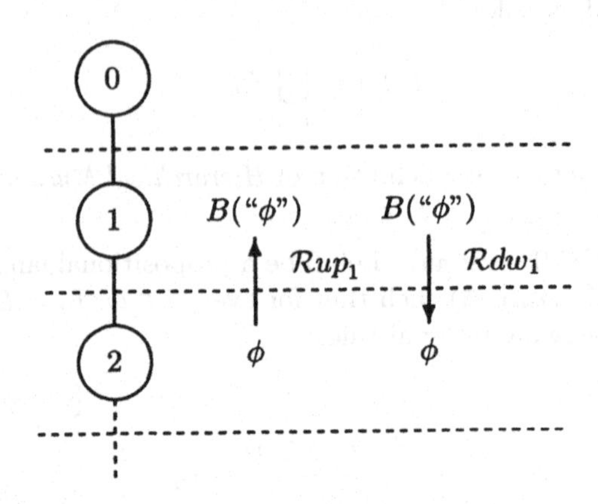

Figure 2. An HMB system

Note that every derivation in $\mathcal{R}upr$ ($\mathcal{R}dwr$) is a derivation in $\mathcal{R}up$ ($\mathcal{R}dw$). We can therefore consider only the following combinations of bridge rules:

$\mathcal{R}up$	$\mathcal{R}dw$	$\mathcal{R}up + \mathcal{R}dw$	$\mathcal{R}up + \mathcal{R}dwr$
$\mathcal{R}upr$	$\mathcal{R}dwr$	$\mathcal{R}upr + \mathcal{R}dwr$	$\mathcal{R}upr + \mathcal{R}dw$

Let us consider theory 0 and theory 1 in Figure 2, that is the theory of a and the view that a has of its own beliefs. In formalizing different reasoning capabilities, bridge rules allow us to obtain different kinds of believers. $\mathcal{R}upr$ makes a complete as it makes it able to prove $B(\text{"}\phi\text{"})$ just because it has proved ϕ in the mental image of its own beliefs. Dually, $\mathcal{R}dwr$ makes a correct. $\mathcal{R}upr + \mathcal{R}dwr$ makes a "ideal", that is, a believes $B(\text{"}\phi\text{"})$ if and only if it believes ϕ in the view that it has of itself. Relaxing the restrictions, from, e.g., $\mathcal{R}dwr$ to $\mathcal{R}dw$, amounts to allowing the application of bridge rules to formulae which are not theorems. This kind of reasoning is very similar to the reasoning informally described in [Haas, 1986; Dinsmore, 1991] and implemented in many applications. These considerations can be trivially generalized to consider a hierarchy of theories of any depth.

REMARK 2. The effect of adding $\mathcal{R}dw$ to $\mathcal{R}upr$ is to obtain an HMB system, called MBK, where the theory of a is theorem equivalent with the minimal normal modal logic $\mathcal{K}$ [Chellas, 1980] (see [Giunchiglia and Serafini, 1994]). Note that $\mathcal{K}$ is the smallest normal system which is meant to model omniscient agents (see [Fagin *et al.*, 1995]).

REMARK 3 (Bounded beliefs: syntax). HMB systems can be easily generalized to formalize bounded beliefs. A bounded HMB system is an HMB system with $I = \{0, 1, \ldots, n\}$ where n is the upper bound on the nesting of the belief predicate B. Notice that each L_i still allows formulae with arbitrary (but finite) depth of nesting of the belief predicate.

3 LOCAL MODELS SEMANTICS

Each view of a is a first-order theory. An "appropriate" semantics should therefore associate to each of them a set of first order models. Bridge rules "link" what holds in the sets of models of the two theories they "connect". Technically bridge rules constrain them to agree on some appropriate sets of theorems (thus, for instance, with $\mathcal{R}upr$, if ϕ holds in all the models of the theory below, then $B(\text{"}\phi\text{"})$ must hold in all the models of the theory above). As for the proof theory, we need to define a formal semantics, called *Local Models Semantics (LMS)* which accounts for both these facts.

The first components of Local Models Semantics are the (local) models of each single theory (and language) in a HMB system.

DEFINITION 4 (Local models). Let $\langle \{L_i\}_{i\in I}, \{\Omega_i\}_{i\in I}, \Delta\rangle$ be an HMB system. For each L_i let a *local model* for L_i be the first order model m for L_i. For each pair (L_i, m) let a *local satisfiability relation* be the (usual) satisfiability relation $\models_{cl}$ between the first order model m and the first order language L_i.

REMARK 5. For each L_i, B is interpreted as a predicate. In particular the above definition says that m satisfies $B(\text{"}\phi\text{"})$ if and only if the interpretation of "ϕ" belongs to the interpretation of B.

Notationally, we write M_i to mean the set of all the local models for L_i.

We can now relate what holds in local models. A *compatibility sequence* $\mathbf{c}$ is a sequence

$$\mathbf{c} = \langle \mathbf{c}_0, \mathbf{c}_1, \ldots, \mathbf{c}_i, \ldots\rangle$$

where, for each $i \in I$, $\mathbf{c}_i$ is a subset of M_i. We call $\mathbf{c}_i$ the i-th element of $\mathbf{c}$. A *compatibility relation* $\mathbf{C}$ is a set of compatibility sequences. Formally, let $\prod_{i\in I} 2^{M_i}$ be the Cartesian product of the collection $\{2^{M_i} : i \in I\}$[3]. The compatibility relation $\mathbf{C}$ is a relation of type

$$\mathbf{C} \subseteq \prod_{i\in I} 2^{M_i}$$

Note that every sequence $\mathbf{c} = \langle \mathbf{c}_0, \mathbf{c}_1, \ldots \mathbf{c}_i, \ldots\rangle$ belonging to a compatibility relation $\mathbf{C}$ is a sequence of sets of local models, being $\mathbf{c}_0$ a set of models of the view a, $\mathbf{c}_1$ a set of models of the view aa, and so on.

[3] Formally the Cartesian product of a collection $\{X_i : i \in I\}$ of sets is denoted by $\prod_{i\in I} X_i$ and is defined to be the set of all functions f with domain I such that $f(i) \in X_i$ for all $i \in I$.

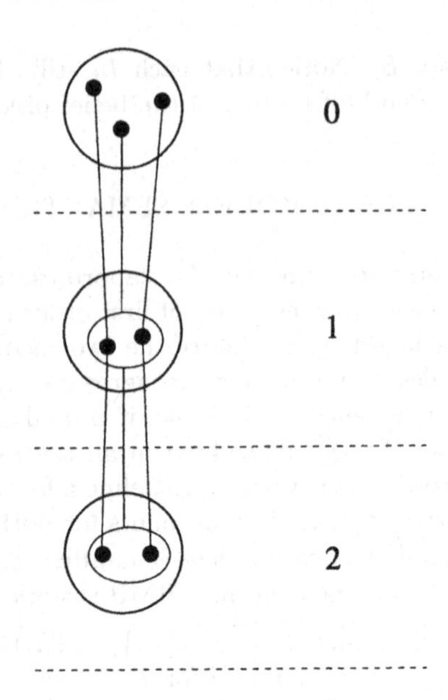

Figure 3. A model for an HMB system

DEFINITION 6 (Model). A *model* for a class $\{L_i\}_{i \in I}$ of languages belonging to an HMB system is a compatibility relation $\mathbf{C}$ satisfying

1. for every sequence $\mathbf{c} = \langle \mathbf{c}_0, \mathbf{c}_1, \dots \mathbf{c}_i, \dots \rangle$ there exists an i with $\mathbf{c}_i \neq \emptyset$
2. for every $m \in \mathbf{c}_i$ there exists $\mathbf{c}' \in \mathbf{C}$ such that $\mathbf{c}'_i = \{m\}$

Condition 1 says that the sequence $\langle \emptyset, \emptyset, \dots, \emptyset, \dots \rangle$ does not belong to the model. Condition 2 imposes the validity of the semantical counterpart of deduction theorem (see Remark 12).

In the following we will write $\mathbf{C}$ to means either a compatibility relation or a model. It will always be clear from the context whether $\mathbf{C}$ is to be interpreted as a compatibility relation or as a model.

Figure 3 gives a graphical representation of the structure of a generic HMB model, where bullets are local models and lines draw the sequences $\langle \mathbf{c}_0, \mathbf{c}_1, \dots \mathbf{c}_i, \dots \rangle$ belonging to a compatibility relation. The set tagged with 0 is the set of models of a, the set tagged with 1 is the set of models representing the view that a has of its own beliefs, and so on. This structure reflects the structure of Figure 2: sets of local models correspond to theories and the relations among sets of local models correspond to bridge rules.

The basic semantic relations of satisfiability, validity and logical consequence are defined as follows:

DEFINITION 7 (Satisfiability). Let $\mathbf{C}$ be a model and $i : \phi$ a formula. $\mathbf{C}$ *satisfies* $i : \phi$, in symbols $\mathbf{C} \models i : \phi$, if for every sequence $\mathbf{c} \in \mathbf{C}$

$$\mathbf{c}_i \models \phi$$

where $\mathbf{c}_i \models \phi$ if every $m \in \mathbf{c}_i$, $m \models_{\mathrm{cl}} \phi$.

REMARK 8. Notice that in the definition of satisfiability we define what it means for a model $\mathbf{C}$ to satisfy a tagged formula $i : \phi$ and not what it means for $\mathbf{C}$ to satisfy ϕ. This takes seriously the idea of having multiple languages.

DEFINITION 9 (Validity). A formula $i : \phi$ is *valid*, in symbols $\models i : \phi$, if every model $\mathbf{C}$ satisfies $i : \phi$.

Defining the logical consequence presents a more subtle problem. It is necessary to take into account that assumptions and conclusion might belong to different languages. For this purpose we use the compatibility relation. The idea is that a formula $i : \phi$ is a consequence of a set of formulae Γ if it is satisfied (at the appropriate level) by every set of local models $\mathbf{c}_i$ that is compatible with all the set of local models $\mathbf{c}_j$ satisfying Γ. Formally:

DEFINITION 10 (Logical consequence). A formula $i : \phi$ is a *logical consequence* of a set of formulae Γ, in symbols $\Gamma \models i : \phi$, if for every model $\mathbf{C}$, every sequence $\mathbf{c}$ in $\mathbf{C}$ satisfies:

(1) if $\mathbf{c} \models \Gamma$ then $\mathbf{c} \models i : \phi$

where $\mathbf{c} \models \Gamma$ means that $\mathbf{c} \models j : \psi$ for every formula $j : \psi$ in Γ.

Let us consider the following example of logical consequence.

EXAMPLE 11. Suppose we want to verify whether $0 : \phi$ is a consequence of $1 : B(\text{“}\phi\text{”})$, i.e. whether $1 : B(\text{“}\phi\text{”}) \models 0 : \phi$, in the class of models containing the two models depicted in Figure 4. We must verify whether condition (1) holds in the two models of the class. The model on the left satisfies (1) because every sequence satisfies $0 : \phi$ (and then (1) is trivially true). The model on the right does not satisfy $1 : B(\text{“}\phi\text{”}) \models 0 : \phi$ because there exists a sequence (thicker line in Figure 4) satisfying $1 : B(\text{“}\phi\text{”})$ and $0 : \neg\phi$.

Notice that Example 11 shows how the compatibility relation allows to propagate consequences (in this case $0 : \phi$ from $1 : B(\text{“}\phi\text{”})$) across different theories.

REMARK 12. It is easy to show that the semantical counterpart of the deduction theorem (i.e. $\Gamma, i : \psi \models i : \phi$ if and only if $\Gamma \models i : \psi \supset \phi$) follows from condition 1 in Definition 6 of model and Definition 10 of consequence relation.

In order to define the classes of models for the HMB systems defined in Section 2 we focus our attention to particular (compatibility) relations

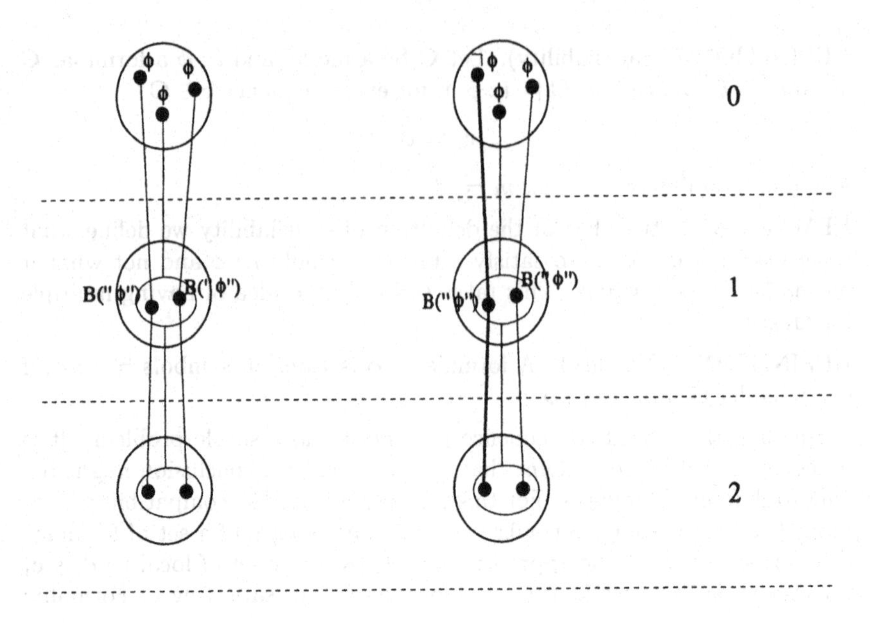

Figure 4. An example of logical consequence

between pairs of adjacent views in Figure 1. We introduce the following notation. Let Γ be a set of L_{i+1}-formulae. We write $B(\text{"}\Gamma\text{"})$ to mean the set of L_i-formulae $B(\text{"}\phi\text{"})$ such that ϕ belongs to Γ. Let Γ be a set of L_i-formulae. We write $B^{-1}(\text{"}\Gamma\text{"})$ to mean the set of L_{i+1}-formulae ϕ such that $B(\text{"}\phi\text{"})$ belongs to Γ.

Let $\mathbf{C}$ be a model. We write $\Theta(\mathbf{c}_i)$ to mean the set of L_i-formulae satisfied by $\mathbf{c}_i$[4]. We write $V^{\uparrow}(\mathbf{c}_i)$ to mean the set of L_{i-1}-formulae satisfied by every sequence $\langle \mathbf{c}'_0, \mathbf{c}'_1, \ldots \mathbf{c}'_i, \ldots \rangle$ in $\mathbf{C}$ such that $\mathbf{c}'_i$ is contained into $\mathbf{c}_i$[5]. Dually, $V^{\downarrow}(\mathbf{c}_i)$ is the set of L_{i+1}-formulae satisfied by every sequence in $\mathbf{C}$ such that $\mathbf{c}'_i$ is contained into $\mathbf{c}_i$[6].

DEFINITION 13 (HMB model). A model $\mathbf{C}$ is an:

(i) $\mathcal{R}up$ model if for every $\mathbf{c} \in \mathbf{C}$ it holds $B(\text{"}\Theta(\mathbf{c}_i)\text{"}) \subseteq \Theta(\mathbf{c}_{i-1})$
(ii) $\mathcal{R}upr$ model if for every $\mathbf{c} \in \mathbf{C}$ it holds $B(\text{"}V^{\downarrow}(\mathbf{c}_i)\text{"}) \subseteq \Theta(\mathbf{c}_i)$
(iii) $\mathcal{R}dw$ model if for every $\mathbf{c} \in \mathbf{C}$ it holds $B^{-1}(\text{"}\Theta(\mathbf{c}_{i-1})\text{"}) \subseteq \Theta(\mathbf{c}_i)$
(iv) $\mathcal{R}dwr$ model if for every $\mathbf{c} \in \mathbf{C}$ it holds $B^{-1}(\text{"}V^{\uparrow}(\mathbf{c}_i)\text{"}) \subseteq \Theta(\mathbf{c}_i)$

[4] Formally $\Theta(\mathbf{c}_i) = \{\phi \in L_i \mid \forall m \in \mathbf{c}_i \; m \models_{cl} \phi\}$.
[5] Formally $V^{\uparrow}(\mathbf{c}_i) = \{\phi \in L_{i-1} \mid \forall \mathbf{c}' \in \mathbf{C} \,.\, \mathbf{c}'_i \subseteq \mathbf{c}_i \Longrightarrow \phi \in \Theta(\mathbf{c}'_{i-1})\}$.
[6] Formally $V^{\downarrow}(\mathbf{c}_i) = \{\phi \in L_{i+1} \mid \forall \mathbf{c}' \in \mathbf{C} \,.\, \mathbf{c}'_i \subseteq \mathbf{c}_i \Longrightarrow \phi \in \Theta(\mathbf{c}'_{i+1})\}$.

For any subset HMB of $\{\mathcal{R}up, \mathcal{R}upr, \mathcal{R}dw, \mathcal{R}dwr\}$, a model **C** is an HMB model, if and only if it is a ζ model for any $\zeta \in$ HMB.

REMARK 14. Notice how we construct the models of combinations of rules (e.g. $\mathcal{R}up + \mathcal{R}dwr$) simply by taking the intersection of the models of the constituent rules (e.g. $\mathcal{R}up$ and $\mathcal{R}dwr$).

REMARK 15. As from remark 2, $\mathcal{R}up + \mathcal{R}dwr$ defines an agent a theorem equivalent to modal $\mathcal{K}$. This semantics naturally allows for a uniform treatment of non normal and normal logics ([Giunchiglia and Giunchiglia, 1996] shows how various non normal modal logics can be captured as appropriate combinations, and restrictions, of the reflections rules introduced above).

Note that every $\mathcal{R}upr$ model is an $\mathcal{R}up$ model and every $\mathcal{R}dwr$ model is an $\mathcal{R}dw$ model. We need therefore to consider only the following HMB models:

$\mathcal{R}up$ model	$\mathcal{R}dw$ model	$\mathcal{R}up + \mathcal{R}dw$ model	$\mathcal{R}up + \mathcal{R}dwr$ model
$\mathcal{R}upr$ model	$\mathcal{R}dwr$ model	$\mathcal{R}upr + \mathcal{R}dwr$ model	$\mathcal{R}upr + \mathcal{R}dw$ model

REMARK 16 (Bounded beliefs: semantics). The semantics for a bounded HMB system is constructed by changing all the definitions of this section to consider $I = \{0, 1, \ldots, n\}$. Thus, for instance, a model is a compatibility relation **C** containing sequences of kind $\langle \mathbf{c}_0, \mathbf{c}_1, \ldots \mathbf{c}_n \rangle$.

4 SOUNDNESS AND COMPLETENESS

In [Giunchiglia and Ghidini, 1997; Ghidini, 1998] we have proved that the the HMB systems of Definition 1 are sound and complete with respect to the classes of HMB models of Definition 13.

THEOREM 17 (Soundness and Completeness). *Let* HMB *be a subset of* $\{\mathcal{R}up, \mathcal{R}dw, \mathcal{R}upr, \mathcal{R}dwr\}$. $\Gamma \vdash_{\text{HMB}} i : \phi$ *if and only if* $\Gamma \models i : \phi$ *holds for every* HMB *model.*

This theorem states that the syntactical notions of derivability and theoremhood in an HMB system are equivalent to semantical notions of logical consequence and validity in the corresponding class of HMB models. Both proofs follow the schema of soundness and completeness theorems for first order logic. In addition we must take into account that derivability and consequence relations may involve formulae belonging to different languages.

The proof of soundness is done by induction on the structure of the derivation. First we prove soundness for propositional rules ($\supset$I, $\supset$E, $\bot$) using condition 2 of Definition 6 of model (see Remark 12). Then we show that conditions i-iv of Definition 13 of HMB model impose soundness of the corresponding bridge rules.

As with the usual canonical model proofs‘of completeness for first order logic, in [Giunchiglia and Ghidini, 1997; Ghidini, 1998] we show that if

$i : \phi$ is not derivable from Γ in an HMB system, then then there exists a (canonical) HMB model $\mathbf{C}^c$ such that $\mathbf{C}^c$ satisfies Γ and $\mathbf{C}^c$ does not satisfy $i : \phi$. The proof of completeness contains the following steps:

1. we generalize, in a natural way, the basic concepts of consistency and maximal consistency (see [Chang and Keisler, 1973]) for a set Γ of formulae of different languages, and we introduce the notions of *k-consistency* and *maximal-k-consistency*. We say that Γ is k-consistent if $k : \bot$ is not derivable from Γ. We say that Γ is maximal-k-consistent if it is k-consistent and the only k-consistent set of formulae containing Γ is Γ itself;

2. we generalize the Lindenbaum's theorem (Lemma 1.2.9 in [Chang and Keisler, 1973]) by showing that for any k-consistent set of formulae Γ there exists a maximal-k-consistent set Γ' with $\Gamma \subseteq \Gamma'$. Moreover Γ' has some nice property similar to the properties of maximal consistent sets in proving completeness for propositional logic;

3. we define the canonical model $\mathbf{C}^c$ as a compatibility relation over sets of (local) models satisfying Γ'. $\mathbf{C}^c$ is an HMB model (i.e. the compatibility relation satisfies conditions of Definition 13), and for every L_k-formula ψ, $\mathbf{C}^c$ satisfies $k : \psi$ if and only if $k : \psi$ belongs to Γ'. This allow us to prove that $\mathbf{C}^c$ satisfies Γ and does not satisfy $i : \phi$.

REMARK 18 (Bounded beliefs: soundness and completeness). The hypothesis that $I = \omega$ does not play any role in booth the proofs of soundness and completeness. Then the same proofs hold when $I = \{0, 1, ..., n\}$.

5 CONCLUSIONS

In this paper we have provided a new semantics for modal logics which captures the intuitions underlying the implementation of propositional attitudes, and belief in particular, inside complex reasoning systems. In the current practice, agents' beliefs and their views about their own or others' beliefs are implemented as belief sets, i.e. as distinct sets of beliefs. Whether an agent has a certain belief is computed by testing whether an "appropriate" formula is a theorem of an "appropriate" belief set, and by "appropriately" propagating this result across belief sets.

These ideas had been already formalized proof-theoretically in the notion of HMB system. This paper provides a semantics which captures the ideas underlying the definition of HMB systems (and, therefore, the current practice in the implementation of belief). Two are the main intuitions underlying the proposed semantics.

- Each first order theory in an HMB system (formalizing a belief set together with the inferential capabilities used to derive consequences from it) defines a set of first order models.
- The propagation of consequences across theories imposes constraints, captured inside (global) compatibility relations, among the (local) models associated to each theory.

ACKNOWLEDGMENTS

This work has benefited from many long discussions with Luciano Serafini and Alex K. Simpson. The collaboration with them has been crucial for the elaboration of the results described in this paper.

Fausto Giunchiglia
IRST, Italy.

Chiara Ghidini
University of Trento, Italy.

BIBLIOGRAPHY

[Chang and Keisler, 1973] C.C. Chang and J.M. Keisler. *Model Theory*. North Holland, 1973.

[Chellas, 1980] B. F. Chellas. *Modal Logic – an Introduction*. Cambridge University Press, 1980.

[Dinsmore, 1991] J. Dinsmore. *Partitioned Representations*. Kluwer Academic Publisher, 1991.

[Fagin *et al.*, 1995] R. Fagin, J.Y. Halpern, Y. Moses, and M. Y. Vardi. *Reasoning about knowledge*. MIT Press, 1995.

[Gabbay, 1990] D. Gabbay. Labeled Deductive Systems. Technical Report CIS-Bericht-90-22, Universität München – Centrum für Informations und Sprachverarbeitung, 1990.

[Gabbay, 1994] D. Gabbay. Labeled Deductive Systems Volume 1 – Foundations. Technical Report MPI-I-94-223, Max-Planck-institut für Informatik, May 1994.

[Ghidini, 1998] C. Ghidini. *A semantics for contextual reasoning: theory and two relevant applications*. PhD thesis, Department of Computer Science, University of Rome "La Sapienza", March 1998. Also IRST-Technical Report 9803-02, IRST, Trento, Italy.

[Giunchiglia and Ghidini, 1997] F. Giunchiglia and C. Ghidini. A Local Models Semantics for Propositional Attitudes. Technical Report 9607-12, IRST, Trento, Italy, 1997.

[Giunchiglia and Ghidini, 1998] F. Giunchiglia and C. Ghidini. Local Models Semantics, or Contextual Reasoning = Locality + Compatibility. In *Proceedings of the Sixth International Conference on Principles of Knowledge Representation and Reasoning (KR'98)*, pp. 282–289. Morgan Kaufmann, 1998. Also IRST-Technical Report 9701-07, IRST, Trento, Italy.

[Giunchiglia and Giunchiglia, 1996] E. Giunchiglia and F. Giunchiglia. Ideal and Real Belief about Belief. In *Practical Reasoning, International Conference on Formal and Applied Practical Reasoning, FAPR'96*, number 1085 in Lecture Notes in Artificial Intelligence, pages 261–275. Springer Verlag, 1996.

[Giunchiglia and Serafini, 1994] F. Giunchiglia and L. Serafini. Multilanguage hierarchical logics (or: how we can do without modal logics). *Artificial Intelligence*, 65:29–70, 1994. Also IRST-Technical Report 9110-07, IRST, Trento, Italy.

[Giunchiglia *et al.*, 1993] F. Giunchiglia, L. Serafini, E. Giunchiglia, and M. Frixione. Non-Omniscient Belief as Context-Based Reasoning. In *Proc. of the 13th International Joint Conference on Artificial Intelligence*, pages 548–554, Chambery, France, 1993. Also IRST-Technical Report 9206-03, IRST, Trento, Italy.

[Giunchiglia, 1993] F. Giunchiglia. Contextual reasoning. *Epistemologia, special issue on I Linguaggi e le Macchine*, XVI:345–364, 1993. Short version in Proceedings IJCAI'93 Workshop on Using Knowledge in its Context, Chambery, France, 1993, pp. 39–49. Also IRST-Technical Report 9211-20, IRST, Trento, Italy.

[Giunchiglia, 1995] F. Giunchiglia. An epistemological science of common sense. *Artificial Intelligence*, 77(2):371–392, September 1995. Also IRST-Technical Report 9503-09, IRST, Trento, Italy.

[Haas, 1986] A. R. Haas. A Syntactic Theory of Belief and Action. *Artificial Intelligence*, 28:245–292, 1986.

[Prawitz, 1965] D. Prawitz. *Natural Deduction - A proof theoretical study*. Almquist and Wiksell, Stockholm, 1965.

LUCIANO SERAFINI AND CHIARA GHIDINI

CONTEXT-BASED SEMANTICS FOR INFORMATION INTEGRATION

1 INTRODUCTION

Due to the increasing necessity and availability of information from different sources, information integration is becoming one of the challenging issues in artificial intelligence and computer science. A successful methodology for information integration is that of federated databases [Sheth and Larson, 1990]. Differently from single databases (for which there are well established formalisms, see for instance [Abitebul *et al.*, 1995]) a completely satisfactory formal treatment of federated databases is still missing (see [Ullman, 1997] for a survey of the state of the art). The goal of this paper is to fill this gap by providing a model theoretic semantics for federated databases.

A federated database is a collection of *distributed*, *redundant*, *partial*, and *partially autonomous* databases which are coordinated by a federated database system. Distribution means that databases of a federated database are different systems, each containing a specific piece of knowledge. Redundancy means that the same piece of knowledge may be represented, possibly from different perspectives, in more than one database. Redundancy not only means that information is duplicated, but also that the information of two databases might be related. Partiality means that the information contained in a database may be incomplete. Autonomy means that each database of a federation has a certain degree of autonomy regarding the design, the execution, and the communication with the other databases. Therefore databases may adopt different conceptual schemata (including domains, relations, naming conventions, ...), and certain operations are performed locally by the databases, without interaction with the other databases.

Distribution, redundancy, partiality, and autonomy generate many problems in the management of a federated database. The most important are: semantic heterogeneity [Sheth and Larson, 1990], interschema dependencies [Catarci and Lenzerini, 1993], query distribution [Arens *et al.*, 1993; Levy *et al.*, 1996], local control over data and processing [Bright *et al.*, 1992], and transparency [Fang *et al.*, 1996]. The definition of a formal semantics for federated databases able to cope with these problems is a key point to understand, specify, and verify the behavior of a federated database. Several approaches have been proposed in the past. An incomplete list is [Catarci and Lenzerini, 1993; Mylopoulos and Motschnig-Pitrik, 1995; McCarthy and Buvac, 1994; Guha, 1990; Blanco *et al.*, 1994; Subrahmanian,

P. Bonzon, M. Cavalcanti and R. Nossum (eds.), Formal Aspects of Context, 175–192.

1994]. However they all fail in representing all these issues in a uniform way. This failure is due, from our perspective, to the fact that these approaches are based on a complete description of the world, and the semantics of the databases are built by filtering the information of such a description. However a description of the real world is hardly to be available, especially in the case of federated databases, which are often constituted by databases developed independently. In most of the cases, indeed, each database in a federation database has its own semantics which corresponds to a partial description of the real world. Therefore the semantics of the federated database must be defined in terms of these partial descriptions. Considering the databases as contexts of a multi contexts system, our intuition is analogous to the semantics of contexts described in [Giunchiglia and Ghidini, 1998]. The semantics for federated databases proposed in this paper, called *Local Model Semantics for federated databases* (LMS hereafter), is an extension to first order language of the semantics of contexts proposed in [Giunchiglia and Ghidini, 1998]. LMS for a federated database is constituted by a set of "local semantics" each formalizing the view of a database, as it was not part of the federation, and by a "compatibility relation" between local semantics, which represents the fact that only certain combinations of views are allowed, as views are on the *same* world. [Serafini and Ghidini, 1997] contains a sound and complete axiomatization of LMS based on Multi Language Systems [Giunchiglia and Serafini, 1994].

The paper is structured as follows. In Section 2 we introduce two motivating examples. In Section 3 we review the basic concepts of semantics for databases and we introduce and motivate LMS for federated databases. In Section 4 we formalize the examples via LMS. Then we compare LMS with the most relevant formalisms for information integration (Section 5), and we make some concluding remarks (Section 6).

2 MOTIVATING EXAMPLES

EXAMPLE 1 (Different Units of Measure). Consider the example of [Sheth and Larson, 1990]. Two schools have the same set of courses, but two different rating systems for their students. Suppose that the students of both schools attend 10 courses and receive a rating from 1 to 10 for each course. At the end of the year the first school assigns a final score to its students based on a scale of six values $\{A, B, C, D, E, F\}$. A student is assigned an A if his total evaluation is less than $\frac{100}{6}$, B if his total is between $\frac{100}{6}$ and $2*\frac{100}{6}$, and so on. The second school instead approximates the final score with a scale of eleven values $\{0, \ldots, 10\}$. The final score of each student is obtained dividing the total evaluation by 10 and rounding to the nearest half point.

Suppose that the two schools want to compare their data. To this purpose

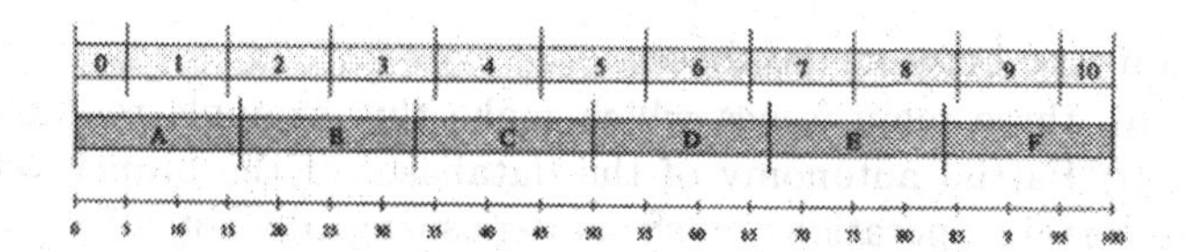

Figure 1. Comparison of the different scales

each database imports the data of the other one. Design autonomy (schools use different scales) imposes transformation of data from a scale to another. The problem is that such a transformation is not trivially definable as a rewriting function. For instance A in the second scale might correspond either to 0 or to 1 or to 2 in the first scale. Figure 1 compares the two scales.

EXAMPLE 2 (Central Bank and Branch Offices). Consider the federated database of a bank with three branch offices. Such a federated database is distributed in four databases: a central database and three databases, one for each branch office. Suppose that each branch office has its own set of customers, and that a customer is allowed to have one or more current accounts in one or more branch offices. The database of each branch office contains the information about its customers, namely the relation between customer names, current account identifiers, and total balances. Let's suppose that this information is confidential and not accessible from the databases of the other branch offices. Only the central bank is allowed to retrieve information about the account balance of each customer from the databases of the branch offices. Branch offices' databases are completely autonomous, there are only the two following constraints: First, current account identifiers are unique. This means that two current accounts, in one or more branch offices, cannot have the same identifier. Second, the sum of current account balances of any customer must be positive. Figure 2 describes the structure

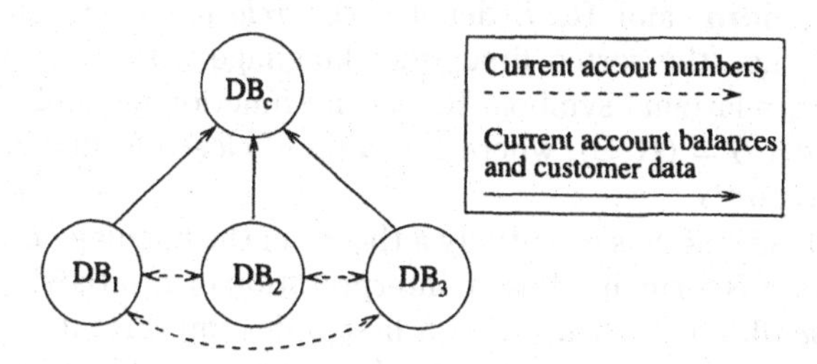

Figure 2. The federated database of the bank example

of the federated database. Circles represent databases, and arrows represent information flow between databases.

There are three main issues which make this example relevant to our purpose. (i) Partial autonomy of the databases of the branch offices implies that certain operations, such as depositing, do not affect the other databases. Therefore these actions must be modelled as local transformations. (ii) Local inconsistency must be modelled. This means that it is possible to have two branch offices with mutually inconsistent data, but this inconsistency must not propagate to the data of the third office. (iii) Global constraints must be respected. For instance, each branch office must be able to prevent a customer to withdraw if (s)he does not have enough money in the whole bank. This must be done interacting with the central bank database, as a branch office does not have information about customer balances in the other branch offices. Also this constraint must be represented in the formal model.

3 LOCAL MODELS SEMANTICS

A natural starting point for the formalization of federated databases is the formalization of their components (the single databases). To this purpose we exploit well established results of this field.

3.1 Technical Preliminaries

We follow [Abitebul *et al.*, 1995] in the notation and terminology. We assume that a countably infinite set **att** of *attributes* is fixed. Let **dom** be a countable set of individual symbols, called *domain*. For any $A \in \mathbf{att}$ the *domain of* A is a non empty subset $dom(A)$ of **dom**. Domains of attributes are mutually disjoint. The set of *relational symbols* is a countable set $\mathbf{R}$ of symbols disjoint from **att** and **dom**, such that for any $R \in \mathbf{R}$, the *sort of* R is a finite sequence of elements of **att**.

Given **att**, **dom**, and $\mathbf{R}$, L denotes the *relational language* over **att**, **dom**, and $\mathbf{R}$, i.e. the sorted first order language with sort **att**, constant symbols **dom**, relational symbols $\mathbf{R}$, and no function symbols. A *database schema* is a pair $S = \langle \mathbf{R}, \Sigma \rangle$, where Σ is a set of closed formulae of L, called *integrity constraints*.

A database schema S is essentially a theory in the language L. A database on the schema S is formalized as an interpretation of L satisfying S. A *complete database* db on a schema S is a first order interpretation of L in the domain **dom**, which maps each $R \in \mathbf{R}$ of sort $\langle A_1, \ldots, A_n \rangle$, into a finite subset of $dom(A_1) \times, \ldots, \times dom(A_n)$, each $d \in \mathbf{dom}$ in itself, and such that db classically satisfies Σ (in symbols db $\models \Sigma$). A complete database contains complete information about the elements of the domain, i.e., for each tuple of elements of the domain and each relation, either the tuple belongs to the relation or not. In many applications, however, it is important to consider databases with partial information, i.e., databases in which it is possible to specify disjunctive facts or existential facts. A ***partial database*** DB on a schema S is a set of complete databases on the schema S. Intuitively a partial (incomplete) database is represented extensionally as the set of all its possible completions. For instance, the partial database in which "John has a car" is the set of interpretations which state that John has a specific car, for any car in the domain of the database. In the following we let the specification "partial" implicit.

An important feature of a database is its query language as we suppose that each database of the federation communicates with the others via query answering. For the purpose of this paper we consider ***first order queries*** (see [Abitebul *et al.*, 1995], chapter 5), i.e. queries defined by first order open formulae. A formula ϕ with free variables in $\{x_1, \ldots, x_n\}$ is denoted by $\phi(x_1, \ldots, x_n)$, and for each tuple $d_1, \ldots, d_n$ of elements of **dom**, The expression obtained by simultaneously replacing each x_i with d_i in ϕ is denoted by $\phi(d_1, \ldots, d_n)$.

3.2 Federated Database Schema

The schema of a federation is composed of the schemata of its components and a schema which describes the redundancy between the schemata of the single databases. Let I be a (at most) countable set of indexes, each of which denotes an element of a federation of databases. The first component of a federated database schema is a family $\{S_i\}_{i \in I}$ (in the following $\{S_i\}$) of schemata of the databases of the federation. We call S_i *local schema.* Let us consider the second component. Databases (as abstractly described here) contain two classes of data, namely objects and relations between objects. With this abstract view, redundancy can be formalized with constraints between the elements of these two classes: constraints on objects and constraints on relations between objects. Constraints on objects, called ***domain constraints***, capture the fact that two databases contain information about a common set of objects of the real world. Constraints on relations, called ***view constraints***, capture the fact that a relation on objects of the real world, represented by a view in a database, is constrained to another relation, represented by another view in another database. Consider Example 2. The set of customers contained in a database of a branch office is contained in the set of customers of the central bank database. Consider Example 1. In the database of both schools there will be a relational symbol $rate(x, y)$ whose intended meaning is the relation which associate to each student his evaluation. Figures 3 a) and b) graphically represent the above examples, respectively.

DEFINITION 3 (Domain Constraint). Let S_i and S_j be two database

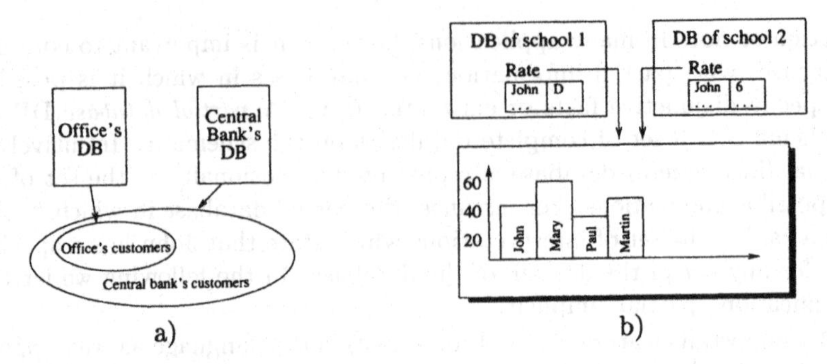

Figure 3. Domain constraints and view constraints

schemata. A *domain constraint from S_i to S_j* is an expression of the form $\mathsf{T}^{i:A}_{j:B}$ or $\mathsf{S}^{i:A}_{j:B}$ where A and B are attributes of S_i and S_j respectively.

Intuitively $\mathsf{T}^{i:A}_{j:B}$ captures the fact that the set of objects of the real world corresponding to attribute A in S_i is contained in the set of objects of the real world corresponding to attribute B in S_j. Conversely $\mathsf{S}^{i:A}_{j:B}$ captures the fact that the set of objects of the real world corresponding to attribute B in S_j is contained in the set of objects of the real world corresponding to attribute A in S_i. Consider Example 1. Suppose that the database schema of both schools contains the attribute *rat-value* with domain $\{A, \ldots, F\}$ in the first school and domain $\{0, \ldots, 10\}$ in the second school. According to how ratings are computed, for each rating value X in $\{A, \ldots, F\}$ there is a rating value Y in $\{0, \ldots, 10\}$ such that the intended meaning of X in the first database (i.e. an integer between 0 and 100) coincides with the intended meaning of Y in the second database. This is formalized by the domain constraint $\mathsf{T}^{1:rat-value}_{2:rat-value}$. Analogously, the fact that for all $X \in \{0, \ldots, 10\}$ there is a $Y \in \{A, \ldots, F\}$ such that the intended meaning of X and Y coincides is represented by $\mathsf{S}^{1:rat-value}_{2:rat-value}$.

Sets of domain constraints from S_i to S_j are denoted by DC_{ij}.

DEFINITION 4 (View Constraint). Let S_i and S_j be two database schemata. A *view constraint from S_i to S_j* is an expression of the form $i : \phi(x_1, \ldots, x_n) \rightarrow j : \psi(x_1, \ldots, x_n)$, where $\phi(x_1, \ldots, x_n)$ and $\psi(x_1, \ldots, x_n)$ are formulae (or equivalently queries) of the language of S_i and S_j, respectively[1].

Intuitively the view constraint $i : \phi(x_1, \ldots, x_n) \rightarrow j : \psi(x_1, \ldots, x_n)$ captures the fact that the relation between real world objects denoted by

[1] We define view constraints as pairs of formulae with the same set of free variables for the sake of simplicity. View constraints can be easily generalized by dropping this requirement.

$\phi(x_1, \ldots, x_n)$ in S_i is contained in the relation between real world objects denoted by $\psi(x_1, \ldots, x_n)$ in S_j. For instance in the example of the two schools the fact that the rating value A in the first database corresponds to 0, 1, or 2, is the second database, is formalized by the view constraint:

$$1 : x = A \rightarrow 2 : x = 0 \vee x = 1 \vee x = 2$$

Sets of view constraints from S_i to S_j are denoted by VC_{ij}. An *interschema constraint* IC_{ij} from S_i to S_j is a pair $IC_{ij} = \langle DC_{ij}, VC_{ij} \rangle$. Frequently occurring domain constraints, such as isomorphism, containment, abstraction, etc. can be expressed by suitable combinations of the domain constraints and view constraints on equality.

The schema for a federated database, called *federated database schema*, is composed of a set of schemata for the databases of the federation, and a set of interschema constraints between pairs of database schemata.

DEFINITION 5 (Federated Database Schema). A *federated database schema on I* is the pair:

$$FS = \langle \{S_i\}, \{IC_{ij}\} \rangle$$

where for each $i \in I$, S_i is a database schema, and for each $j \in I$ with $j \neq i$, IC_{ij} is an interschema constraint from S_i to S_j.

3.3 Federated Database

The aim of a federated database schema is the formal specification of a class of federated databases. In this section we formally define a federated database and when a federated database is of a given federated database schema.

Being a federated database composed of a set of distributed and autonomous databases, its formal model must be specifiable in terms of the composition of the models of each single database. We take the perspective described in [Giunchiglia and Ghidini, 1998] for the semantics of contexts by formalizing each database as a context. The formal semantics associated to each database i represents the description of the real world from the i-th partial point of view. Therefore the formal semantics of a federated database on the federated database schema $\langle \{S_i\}, \{IC_{ij}\} \rangle$ contains a set $\{\mathrm{DB}_i\}$ of databases, each DB_i being a partial database on the schema S_i.

According to this perspective, the databases of a federation may have distinct domains, namely there is no global domain associated to the federated database, there is rather a set of domains, each associated to a single database. Let $\mathbf{dom}_i$ be the domain of DB_i. The fact that two object in $\mathbf{dom}_i$ and $\mathbf{dom}_j$ formalize the same object of the real world is represented by a relation $r_{ij} \subseteq \mathbf{dom}_i \times \mathbf{dom}_j$ between the domains of the two databases. A pair $\langle d, d' \rangle$ being in r_{ij} means that d in the first database and d' in the second one represent the same object of the real world. r_{ij} is called *domain relation from i to j*.

A possible domain relation of Example 1, restricted to the domains of rating values, is the following:

$$r_{12} = \left\{ \begin{array}{cccccc} \langle 0, A \rangle, & \langle 1, A \rangle, & \langle 2, B \rangle, & \langle 3, B \rangle, & \langle 3, C \rangle, & \langle 4, C \rangle, \\ \langle 5, D \rangle, & \langle 6, D \rangle, & \langle 7, E \rangle, & \langle 8, F \rangle, & \langle 9, F \rangle, & \langle 10, F \rangle \end{array} \right\}$$

A formal semantics of a federated database is composed of a set of databases (as defined in Section 3.1) and a set of domain relations from the schema of a component to that of the others. Domain constraints imply that only certain domain relations are accepted. Analogously view constraints imply that only certain combinations of databases are admitted.

DEFINITION 6 (Satisfiability of Domain Constraint). Let S_i and S_j be two database schemata. The domain relation r_{ij} satisfies the domain constraint $\mathsf{T}^{i:A}_{j:B}$ iff for any $d \in dom_i(A)$ there is a $d' \in dom_j(B)$ such that $\langle d, d' \rangle \in r_{ij}$. Analogously r_{ij} satisfies the domain constraint $\mathsf{S}^{i:A}_{j:B}$ iff for any $d \in dom_j(B)$ there is a $d' \in dom_i(A)$ such that $\langle d', d \rangle \in r_{ij}$.

DEFINITION 7 (Satisfiability of View Constraint). Let DB_i and DB_j be two databases on the schema S_i and S_j, and r_{ij} be a domain relation. The tuple $\langle \mathrm{DB}_i, \mathrm{DB}_j, r_{ij} \rangle$ satisfies the view constraint $i : \phi(x_1, \ldots, x_n) \to j : \psi(x_1, \ldots, x_n)$ iff for any $\langle d_k, d'_k \rangle \in r_{ij}$ $(1 \leq k \leq n)$, $\mathrm{DB}_i \models \phi(d_1, \ldots, d_n)$ implies that $\mathrm{DB}_j \models \psi(d'_1, \ldots, d'_n)$.

An intuitive interpretation of satisfiability of a view constraint from DB_i to DB_j (Definition 7), can be given in terms of relations between the results of queries to the databases. A domain relation r_{ij} can be interpreted as a mapping from relations in i into relations in j. Formally if $X \subseteq \mathbf{dom}_i^n$ is a relation in DB_i then $r_{ij}(X)$ is defined as

$$\{\langle d'_1, \ldots, d'_n \rangle \in \mathbf{dom}_j^n \mid \langle d_1, \ldots, d_n \rangle \in X \text{ and for all } 1 \leq k \leq n \ \langle d_k, d'_k \rangle \in r_{ij}\}$$

According to this definition, $\langle \mathrm{DB}_i, \mathrm{DB}_j, r_{ij} \rangle$ satisfies the view constraint $i : \phi(x_1, \ldots, x_n) \to j : \psi(x_1, \ldots, x_n)$ if and only if $r_{ij}(X) \subseteq Y$, being X and Y the result of the query $\phi(x_1, \ldots, x_n)$ to DB_i, and $\psi(x_1, \ldots, x_n)$ to DB_j respectively.

DEFINITION 8 (Federated Database). Let $\{S_i\}$ be a set of database schemata. Let $\{\mathrm{DB}_i\}$ be a set of databases, each DB_i being a database on S_i, and $\{r_{ij}\}$ be a family of domain relations. A *federated database* on the federated database schema $\langle \{S_i\}, \{IC_{ij}\} \rangle$ is a pair

$$FDB = \langle \{\mathrm{DB}_i\}, \{r_{ij}\} \rangle$$

such that for all i, j, $i \neq j$, $\langle \mathrm{DB}_i, \mathrm{DB}_j, r_{ij} \rangle$ satisfies IC_{ij}.

3.4 Logical Consequence

Interschema constraints, like integrity constraints in single databases (see for instance [Abitebul *et al.*, 1995] Chapter 8), imply that certain facts in a database are consequences of other facts in, possibly distinct, databases. The formal characterization of such a relation is crucial as it allows to formally check inconsistencies in the databases and to understand how information propagates through databases. In this section we formalize this relation between facts by the notion of *logical consequence* in a federated database (or more simply logical consequence). Logical consequence is a relation between formulae of the relational languages of the databases. To define logical consequence we introduce some extra notation. A *labelled formula* is a pair $i : \phi$. It denotes the formula ϕ and the fact that ϕ is a formula of the database schema S_i. If no ambiguity arises, labelled formulae are called formulae. Given a set of labelled formulae Γ, Γ_j denotes the set of formulae $\{\gamma \mid j : \gamma \in \Gamma\}$. From now on we say that ϕ is a *i-formula* to specify that ϕ is a formula of the database schema S_i.

We extend the set of variables of each S_i of a federated database to a set of *extended variables*. For each $i, j \in I$, each variable x of sort A in S_i, and each attribute B in S_j, $x^{j:B\rightarrow}$ and $x^{\rightarrow j:B}$ are variables of sort A.

Intuitively a variable x of sort A (without indexes) occurring in $i : \phi$ is a placeholder for a generic element of $dom_i(A)$; the extended variable $x^{j:B\rightarrow}$ of sort A occurring in $i : \phi$ is a placeholder for an element of $dom_i(A)$ which is an image, via the domain relation r_{ji}, of the element of $dom_j(B)$ denoted by x; analogously $x^{\rightarrow j:B}$ occurring in $i : \phi$ is a placeholder for an element of $dom_i(A)$ which is a pre-image, via r_{ij}, of the element of $dom_j(B)$ denoted by x.

Given a federated database FDB on a federated database schema FS, an *assignment* is a total function a which maps a pair $\langle e, i\rangle$, where e is an extended variable (namely x, or $x^{j:B\rightarrow}$, or $x^{\rightarrow j:B}$) of the language of S_i into an element of $dom_i(A)$. a is an *admissible assignment* if and only if for any variable x of sort A and any variable $x^{i:A\rightarrow}$ and $x^{\rightarrow i:A}$ of sort B

1. if $\mathsf{T}^{i:A}_{j:B} \in DC_{ij}$, then $\langle a(x,i), a(x^{i:A\rightarrow}, j)\rangle \in r_{ij}$

2. if $\mathsf{S}^{i:B}_{j:A} \in DC_{ji}$, then $\langle a(x^{\rightarrow i:A}, j), a(x,i)\rangle \in r_{ji}$.

DEFINITION 9 (Satisfiability). Let $FDB = \langle\{\mathrm{DB}_i\}, \{r_{ij}\}\rangle$ be a federated database. A i-formula ϕ is satisfied in db $\in \mathrm{DB}_i$ by an assignment a, in symbols db $\models \phi[a]$, according to the definition of satisfiability of first order logic. A i-formula ϕ is satisfied in DB_i by an assignment a, in symbols $\mathrm{DB}_i \models \phi[a]$, if for any db $\in \mathrm{DB}_i$, db $\models \phi[a]$. $i : \phi$ is satisfied in FDB by an assignment a, in symbols $FDB \models i : \phi[a]$, if $\mathrm{DB}_i \models \phi[a]$.

Notationally for any set of i-formulae Γ_i, $\mathrm{DB}_i \models \Gamma_i[a]$ means that $\mathrm{DB}_i \models \gamma[a]$ for any $\gamma \in \Gamma_i$. Let DB_i be a database on S_i and a an assignment for any set of i-formulae Γ, ϕ. $\Gamma[a] \models_{\mathrm{DB}_i} \phi[a]$ if and only if for all $\mathrm{db}_i \in \mathrm{DB}_i$, $\mathrm{db}_i \models \Gamma[a]$ implies that $\mathrm{db}_i \models \phi[a]$.

DEFINITION 10 (Logical Consequence). Let FS be a federated database schema. Let Γ be a set of formulae. A formula $i : \phi$ is a *logical consequence* of Γ, in symbols $\Gamma \models_{FS} i : \phi$, iff for any federated database $\langle\{\mathrm{DB}_i\}, \{r_{ij}\}\rangle$ on FS and for any admissible assignment a, if for all $j \neq i$, $\mathrm{DB}_j \models \Gamma_j[a]$, then $\Gamma_i[a] \models_{\mathrm{DB}_i} \phi[a]$.

4 MODELING THE EXAMPLES

EXAMPLE 11 (Formalization of Example 1). The scenario of Example 1 is representable by two databases, which describe at a different approximation level, a world of students having rating numbers between 1 and 100.

Local Schemata Let S_1 and S_2 be the database schemata for the two schools. S_1 and S_2 contain two attributes *rat-value* (for rating values) and *student* (for students). $dom_1(\textit{rat-value}) = \{A, \ldots, F\}$ and $dom_2(\textit{rat-value}) = \{1, \ldots, 10\}$. $dom_1(\textit{student})$ and $dom_2(\textit{student})$ are the set of proper names. S_1 and S_2 contain also a binary predicate $rate(x, y)$ of sort $\langle \textit{student}, \textit{rat-value}\rangle$ meaning that the final rate of student x is y.

Interschema Constraints Let's first consider domain constraints. The two schools represent, via the attribute *student*, the same set of students. This is formalized by the domain constraints:

$$(1) \quad \mathsf{T}^{1:student}_{2:student} \qquad \mathsf{S}^{1:student}_{2:student} \qquad \mathsf{T}^{2:student}_{1:student} \qquad \mathsf{S}^{2:student}_{1:student}$$

Both schools represent, via the attribute *rat-value*, the same set of rating values (i.e. the set of positive integers ≤ 100) at different levels of approximation. This is formalized by the domain constraints analogous to (1) on the attribute *rat-value*.

Let's consider view constraints. Both schools agree on students' names. E.g. the intended meaning of "John" in the database of both schools is a unique person whose name is John. This is represented by the view constraints:

$$(2) \quad 1 : x = c \to 2 : x = c \qquad\qquad 2 : x = c \to 1 : x = c$$

for any student name c. Rating transformation is formalized by two sets of view constraints that reflect the comparison between the two different scales in Figure 1:

$$\begin{array}{ll} 1 : x = A \to 2 : x = 0 \vee x = 1 \vee x = 2 & 2 : x = 0 \to 1 : x = A \\ 1 : x = B \to 2 : x = 2 \vee x = 3 & 2 : x = 1 \to 1 : x = A \\ \vdots & \vdots \\ 1 : x = F \to 2 : x = 8 \vee x = 9 \vee x = 10 & 2 : x = 10 \to 1 : x = F \end{array}$$

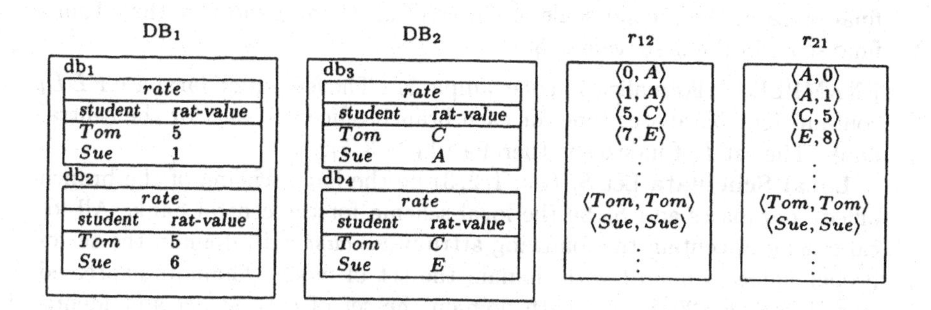

Figure 4. A federated database on FS_s

Finally, the intended meaning of the predicate $rate(x, y)$ in both databases coincides. This is formalized by the view constraints:

(3) $1 : rate(x, y) \rightarrow 2 : rate(x, y) \qquad 2 : rate(x, y) \rightarrow 1 : rate(x, y)$

The federated database schema formalizing this example is therefore $FS_s = \langle\{S_1, S_2\}, \{IC_{12}, IC_{21}\}\rangle$ where IC_{12} and IC_{21} contain the domain constraints and view constraints defined above.

An example of federated database FDB on the schema FS_s is shown in Figure 4. In FDB, $\mathrm{DB}_1 \models rate(Tom, 5)$ and, by view constraint (3), $\langle 5, C\rangle \in r_{12}$ forces $\mathrm{DB}_2 \models rate(Tom, C)$. Another federated database on FS_s is obtainable form FDB by replacing $\langle 5, C\rangle$ with $\langle 5, D\rangle$ in r_{12}. Again view constraint (3) forces $\mathrm{DB}_2 \models rate(Tom, D)$. However, in order to satisfy view constraint (2) and domain constraint $\mathrm{T}^{1:rat-value}_{2:rat-value}$, either $\langle 5, C\rangle \in r_{12}$, or $\langle 5, D\rangle \in r_{12}$. This implies that, for any federated database on FS_s, $\mathrm{DB}_1 \models rate(Tom, 5)$ implies $\mathrm{DB}_2 \models rate(Tom, C) \vee rate(Tom, D)$. The above observations are summarized by the following properties of the logical consequence of FS_s:

(4) $2 : rate(Tom, 5) \not\models_{FS_s} 1 : rate(Tom, C)$

(5) $2 : rate(Tom, 5) \not\models_{FS_s} 1 : rate(Tom, D)$

(6) $2 : rate(Tom, 5) \models_{FS_s} 1 : rate(Tom, C) \vee rate(Tom, D)$

Notice that the properties of $\models_{FS_s}$ shown above formalize that semantic heterogeneity between the two scales prevents to find a one to one translation between rates. In particular the fact that neither $1 : rate(Tom, C)$ nor $1 : rate(Tom, D)$ are logical consequences of $2 : rate(Tom, 5)$ (equations (4) and (5)) formalizes that we cannot translate the rate 5 to a unique value (C or D) because of the fact that 5 might be obtained rounding off a valuation between 4.5 and 5, or by rounding off a valuation between 5 and 5.5. However equation (6) enable us to infer the partial information that Tom's final score in the second scale is either C or D from the fact that Tom's final score in the first scale is 5.

EXAMPLE 12 (Formalization of example 2). The federated database FDB_b contains four databases: one for each branch office and one for the central bank. The set I of indexes is therefore $\{1, 2, 3, b\}$.

Local Schemata Let S_i $(i = 1, 2, 3)$ be the local schema of the branch offices' databases and S_b be the local schema for the central bank. All local schemata contain the following attributes: *cust* with domain the set of proper names; *amount* with domain the set of money amounts expressed in dollars (e.g. 200\$); *cai* with domain the set of current account identifiers. Each S_i $(i = 1, 2, 3)$ contains the predicate $account(x, y, z)$ of sort $\langle cust, cai, amount\rangle$ meaning that current account y of customer x has final balance z. S_b contains three predicates $balance1(x, y)$, $balance2(x, y)$, and $balance3(x, y)$ of sort $\langle cust, amount\rangle$ meaning that the total balance of the current accounts of customer x in the first (resp. second and third) branch office is y.

For the sake of the example let us suppose that the total balance of the current accounts of a customer in all branch offices must be positive. This is formalized by the following integrity constraint[2]

$$(7) \quad balance1(x, y) \wedge balance2(x, y') \wedge balance3(x, y'') \supset y + y' + y'' > 0$$

Interschema Constraints Let us start with domain constraints. The set of customers of each branch office is contained in the set of customer of the central bank; this is formalized by the domain constraint $\mathsf{T}^{i:cust}_{b:cust}$ The currency used in all offices is unique; this is formalized by the domain constraints $\mathsf{T}^{i:amount}_{j:amount}$, and $\mathsf{S}^{i:amount}_{j:amount}$ for any $i, j = 1, 2, 3, b$. Let us now consider view constraints. The central bank is allowed to access the information concerning the customers' balance. This is represented by the following view constraints

$$(8) \qquad 1 : \exists^n y_i \; account(x, y_i, z_i) \rightarrow b : balance1(x, \textstyle\sum_i z_i)$$

$$(9) \qquad 2 : \exists^n y_i \; account(x, y_i, z_i) \rightarrow b : balance2(x, \textstyle\sum_i z_i)$$

$$(10) \qquad 3 : \exists^n y_i \; account(x, y_i, z_i) \rightarrow b : balance3(x, \textstyle\sum_i z_i)$$

where for any $n \geq 0$, $\exists^n y_i \phi(x, y_i, z_i)$ means "there exists exactly n distinct individuals $y_1, \ldots, y_n$ such that $\phi(x, y_1, z_1) \wedge \ldots \wedge \phi(x, y_n, z_n)$"[3]. Constants

[2] To simplify the notation, we use the infix functional symbol $+$. Equivalence formulas can be written by translating $+$ in a relational symbol and by adding suitable constraints.

[3] Formally $\exists^n y_i \phi(x, y_i, z_i)$ is defined as

$$\exists y_1 \ldots y_n \left(\bigwedge_{i=1}^{n} \phi(x, y_i, z_i) \wedge \bigwedge_{i \neq j = 1}^{n} y_i \neq y_j \wedge \forall y' z' \left(\phi(x, y', z') \supset \bigvee_i y' = y_i \right) \right)$$

for customers, current account identifiers, and amounts expressed in dollars have the same meaning in all the databases. This is formalized by the view constraint (2) from any database schema to the others, where c is either a customer name, a current account identifier, or amount in dollars. The requirement that account identifiers are unique in all the branch offices is formalized by the following view constraints, for any current account identifier c of the language of DB_i

$$(11) \quad i : \exists xz.account(x, c, z) \rightarrow j : \forall xz.\neg account(x, c, z) \text{ for any } i, j = 1, 2, 3$$

The federated database schema formalizing this example is a pair $FS_b = \langle\{S_i\}, \{IC_{ij}\}\rangle$, where $I = \{1, 2, 3, b\}$, and S_i and IC_{ij} are the local schemata and the interschema constraints described above. A model is a pair $FDB_b = \langle\{\mathrm{DB}_1, \mathrm{DB}_2. \mathrm{DB}_3, \mathrm{DB}_b\}, \{r_{ij}\}\rangle$ on FS_b.

Let's make some remarks on how the models which satisfies the above interschema constraints address the issues described in Example 2. (i) Suppose that DB_1 contains $account(John, C45, 200\$)$, and that John deposits 100$ in current account $C45$. This operation forces DB_1 to be updated into DB_1' by replacing $account(John, C45, 200\$)$ with $account(John, C45, 300\$)$. In order to keep the federated database this change must be propagated following the interschema constraints. Therefore DB_b must be updated accordingly. obtaining DB_b'. Notice that DB_2 and DB_3 need not to be updated. as constraints between branch offices database does not allow to export information about balances to other databases. (ii) Suppose that the first and the second branch office assign the same account identifier $C3$ to *John* and *Richard*. i.e., suppose that $\mathrm{DB}_1 \models account(John, C3, 100\$)$ and $\mathrm{DB}_2 \models account(Richard, C3, 200\$)$. By the domain constraints and view constraints it follows that

$$(12) \quad 1 : account(John, C3, 100\$),\ 2 : account(Richard, C3, 200\$) \models_{FS_b} 1 : \bot$$

$$(13) \quad 1 : account(John, C3, 100\$),\ 2 : account(Richard, C3, 200\$) \models_{FS_b} 2 : \bot$$

Notice that the following equation ensures that inconsistency is not propagated to the other branch offices

$$(14) \quad 1 : account(John, C3, 100\$)\ 2 : account(Richard, C3, 200\$) \not\models_{FS_b} 3 : \bot$$

(iii) Each branch office may prevent a customer to withdraw if this action violates the integrity constraint (7) in the central bank. Suppose that *Richard* has three accounts one in each branch office:

$$\begin{aligned} \mathrm{DB}_1 &\models account(Richard, C3, 100\$) \\ \mathrm{DB}_2 &\models account(Richard, F5, 200\$) \\ \mathrm{DB}_3 &\models account(Richard, HT65, 150\$) \end{aligned}$$

Suppose that *Richard* tries to withdraw 500\$ from current account C3 in the first branch office. According to how we model deposit and withdraw operations, we should update FDB by substituting $account(Richard, C3, 100\$)$ with $account(Richard, C3, -400\$)$ in DB_1, and by updating DB_b accordingly, obtaining a database DB'_b (DB_2 and DB_3 doesn't change). Due to the interschema constraints we can infer that:

$$\begin{array}{l} \mathrm{DB}'_b \models balance1(Richard, -400\$) \\ \mathrm{DB}'_b \models balance2(Richard, 200\$) \\ \mathrm{DB}'_b \models balance3(Richard, 150\$) \end{array}$$

Because of the integrity constraint (7) DB'_b is the empty set; i.e. DB'_b is inconsistent. Thus DB_1 is not allowed to modify the interpretation of *account* and *Richard* cannot withdraw.

5 RELATED WORK

A significant attempt to develop a logic based formal semantics for heterogeneous information integration is the idea of *cooperative information system* (CIS) described in [Catarci and Lenzerini, 1993]. A CIS is quite similar to a federated databases. It is composed of a set of database schemata and a set of so called interschema assertions. Database schemata represent the individual information sources and are theories in description logics [Borgida, 1995]. Interschema assertions formalize relations between different database schemata. CISs formalize a certain degree of autonomy, each database having its own language, domain, and schema. Furthermore CISs formalize a certain degree of redundancy by means of interschema assertions, which capture four different kinds of semantic interdependencies between concepts and relations in different databases. A first difference between CIS and LMS concerns the domains. A model for a CIS is defined over a global domain which is the union of the domains of the databases. This implies that a constant c in different databases is interpreted in the same object c in the CIS. As a consequence in CIS one cannot represent various forms of redundancy between objects belonging to different database domains, e.g. the fact that a database domain is an abstraction of another database domain. A second difference concerns partiality. CIS models complete databases and cannot express partiality. Totality affects directionality. Indeed in CIS every interschema constraints from S_1 to S_2 entails the converse interschema assertion in the opposite direction. This prevents CIS to completely represent directionality in the communication between databases.

Blanco *et al.* [1994] have developed a formalism similar to [Catarci and Lenzerini, 1993] which suffers from similar problems.

Subrahmanian [1994; 1996] uses annotated logic [Blair and Subrahmanian, 1987] to integrate a set of deductive databases in an unique amalgamated database called amalgam. The amalgam, in addition to the disjoint union of the databases, contains a supervisory database. The supervisory database is a set of clauses (called amalgamated clauses) which resolve conflicts due to inconsistent facts and compose information of different database sources. [Subrahmanian, 1994] investigates the relation between the models of the amalgam and the models of its components. Subrahmanian takes a more general approach then ourselves as he considers formulas with complex sets of truth values and time intervals. However the intuition behind amalgamated clauses (contained in the supervisory database) is very close to that of a generalization of view constraints described in [Ghidini and Serafini, 1997]. From our perspective adopting a global amalgamated database is the reason of the main drawback of Subrahmanian's approach. Indeed global amalgamated database prevents one to associate distinct deductive mechanism to each database in the federation. Furthermore amalgamated database doesn't support local inconsistency. I.e. the inconsistency of a local database forces the inconsistency of the whole amalgamated database. Differently, our approach allows federated databases in which the i-th database is inconsistent while the other are consistent.

Vermeer *et al.* [1996] exploit the semantic information provided by the integrity constraints of the single databases to achieve interoperability among them. The spirit of this approach is similar to ours, although they mainly address a different problem. In [Vermeer and Apers, 1996] different databases are integrated in an unique integrated view. The consistency of such an integrated view is checked by using integrity constraints of component databases. Vermeer *et al.* argue that semantic relations are expressed by relationships between objects (cf. domain relations) and relations between classes are the result of object relationships (cf. definition of satisfiability of domain/view constraints w.r.t. a domain relation).

A context-based approach to the problem of specifying redundancy between different databases, maintaining an hight degree of autonomy has been proposed by Mylopoulos *et al.* [1995]. Mylopoulos *et al.* do not address the issue of information integration, they rather describe a set of criteria for splitting a database in a set of (possibly overlapping) partitions. However this work is relevant here as it provides mechanisms for the management of different overlapping partitions. In this work partitions are represented as contexts, exploiting a notion of context very closed to that described in [Giunchiglia, 1993]. Contexts represent the fact that partitions may differ in their viewpoints, focus of attention, topics, and history. Moreover contexts allow to associate specific languages and semantics to each partition. [Mylopoulos and Motschnig-Pitrik, 1995] does not provide any formal semantics. However we think that LMS would be a good candidate to formalize most of the static issues addressed in this work.

6 CONCLUSIONS

In this paper we have provided a context based semantics, called *Local Models Semantics for federated databases.* We have shown that this LMS addresses some of the relevant problems in heterogeneous information integration, such as semantic heterogeneity, interschema dependencies, local control over data and processing and transparency. The major contributions of this paper are the following: First we have described motivating examples in federated databases and within each example we have pointed out its critical aspects. These aspects cover a large range of well known problems in the area of federated databases. Second we have provided a context based semantics for federated databases. We have introduced the key concepts of domain constraint and view constraint. Domain constraints and view constraints formalize two orthogonal aspects of relation between databases. Domain constraints formalize overlapping between domains of databases; view constraints formalize heterogeneity and redundancy of database schemata, namely relations between symbols of database schemata. Third we have defined a notion of logical consequence between formulae (queries) in different databases. This notion constitutes the theoretical ground for the implementation of correct algorithm for query answering, query optimization, schema consistency checking, and schema reduction. Forth we have argued that LMS is an adequate formalism for federated databases by formalizing the motivating examples presented in the paper. Finally we have compared LMS with the main formalisms in the area of information integration. We have provided (but not described in this paper) a context based proof theory [Serafini and Ghidini, 1997] for LMS. Future work is to develop an algorithms for query answering, query optimization, schema consistency checking, and schema reduction, and prove their correctness w.r.t LMS.

ACKNOWLEDGMENTS

The authors would like to thank Fausto Giunchiglia for encouragement, basic ideas, discussions on motivations, and Alessandro Artale. Massimo Benerecetti, Camilla Casotto, Alessandro Cimatti for useful feedback and discussions. The authors also thank all the members of the Mechanized Reasoning Group at DISA (University of Trento), DIST (University of Genoa) and ITC-IRST (Trento). This work is part of the MRG project on *Distributed Representations and Systems.*

Luciano Serafini
ITC-IRST, Trento, Italy.

Chiara Ghidini
University of Trento, Italy.

BIBLIOGRAPHY

[Abitebul *et al.*, 1995] S. Abitebul, R. Hull, and V. Vianu. *Foundation of Databases*. Addison-Wesley, 1995.

[Adali and Subrahmanian, 1996] Sibel Adali and V.S. Subrahmanian. Amalgamating Knowledge Bases, III: Algorithms, Data Structures, and Query Processing. *Journal of Logic Programming*, 28(1):45–88, 1996.

[Arens *et al.*, 1993] V. Arens, C.Y. Chee, C-N. Hsu, and C.A. Knoblock. Retrieving and integrating data from multiple information sources. *International Journal on Intelligent and Cooperative Information Systems*, 2(2):127–158, 1993.

[Blair and Subrahmanian, 1987] H.A. Blair and V.S. Subrahmanian. Paraconsistent Logic Programming. *Theoretical Computer Science*, 68:35–51, 1987.

[Blanco *et al.*, 1994] J.M. Blanco, A. Illarramendi, and A. Goñi. Building a federated relational database system: an approach using a knowledge-based system. *International Journal of Intelligent and Cooperative Information Systems*, 3(4):415–455, 1994.

[Borgida, 1995] A. Borgida. Description Logics in Data Management. *IEEE Transactions on Knowledge and Data Engineering*, October 1995.

[Bright *et al.*, 1992] M.W. Bright, A. R. Hurson, and Simin H. Pakzad. A Taxonomy and Current Issues in Multidatabase Systems. *Computer*, 25(3):50–60, March 1992.

[Catarci and Lenzerini, 1993] T. Catarci and M. Lenzerini. Representing and using interschema knowledge in cooperative information systems. *International Journal of Intelligent and Cooperative Information Systems*, 2(4):375–398, 1993.

[Fang *et al.*, 1996] D. Fang, S. Ghandeharizadeh, and D. McLeod. An experimental object-based sharing system for networked databases. *The VLDB Journal*, 5:151–165, 1996.

[Ghidini and Serafini, 1997] C. Ghidini and L. Serafini. Distributed First Order Logics. In *Proceedings of the Second International Workshop on Frontiers of Combining Systems (FroCoS'98)*, Amsterdam, Holland, 1998. Also IRST-Technical Report 9709-02, IRST, Trento, Italy.

[Giunchiglia and Ghidini, 1998] F. Giunchiglia and C. Ghidini. Local Models Semantics, or Contextual Reasoning = Locality + Compatibility. In *Proceedings of the Sixth International Conference on Principles of Knowledge Representation and Reasoning (KR'98)*, pp. 282–289. Morgan Kaufmann, 1998. Also IRST-Technical Report 9701-07, IRST, Trento, Italy.

[Giunchiglia and Serafini, 1994] F. Giunchiglia and L. Serafini. Multilanguage hierarchical logics (or: how we can do without modal logics). *Artificial Intelligence*, 65:29–70, 1994. Also IRST-Technical Report 9110-07, IRST, Trento, Italy.

[Giunchiglia, 1993] F. Giunchiglia. Contextual reasoning. *Epistemologia, special issue on I Linguaggi e le Macchine*, XVI:345–364, 1993. Short version in Proceedings IJCAI'93 Workshop on Using Knowledge in its Context, Chambery, France, 1993, pp. 39–49. Also IRST-Technical Report 9211-20, IRST, Trento, Italy.

[Guha, 1990] R.V. Guha. Microtheories and contexts in cyc. Technical Report ACT-CYC-129-90, MCC, Austin, Texas, 1990.

[Levy *et al.*, 1996] A.Y. Levy, A. Rajaraman, and J.J. Ordille. Querying Heterogeneous Information Sources Using Source Descriptions. In *Proceedings of the 22nd VLDB Conference*, Bombay, India, 1996.

[McCarthy and Buvac, 1994] J. McCarthy and S. Buvac. Notes on Formalizing Context (Expanded Notes). Technical Report CS-TN-94-13, Stanford University, Stanford, CA, 1994.

[Mylopoulos and Motschnig-Pitrik, 1995] J. Mylopoulos and R. Motschnig-Pitrik. Partitioning Information Bases with Contexts. In *Third International Conference on Cooperative Information Systems*, Vienna, 1995.

[Serafini and Ghidini, 1997] L. Serafini and C. Ghidini. Formalizing and reasoning about constraints in federated databases. Technical Report 9704-01, IRST, Trento, Italy, April 1997.

[Sheth and Larson, 1990] A. Sheth and J. Larson. Federated database systems for managing distributed, heterogeneous, and autonomous databases. *ACM Computing Surveys*, 22(3):183–236, 1990.

[Subrahmanian, 1994] V.S. Subrahmanian. Amalgamating Knowledge Bases. *ACM Trans. Database Syst.*, 19(2):291–331, 1994.

[Ullman, 1997] Jeffrey D. Ullman. Information Integration Using Logical Views. In *Proc. of the 6th International Conference on Database Theory (ICDT'97)*, 1997.

[Vermeer and Apers, 1996] M.W.W. Vermeer and P.M.G. Apers. The Role of Integrity Constraints in Database Interoperation. In *Proceedings of the 22nd VLDB Conference*, Mumbai(Bombay), India, 1996.

DOV M. GABBAY AND ROLF T. NOSSUM

STRUCTURED CONTEXTS WITH FIBRED SEMANTICS

1 INTRODUCTION

The logical treatment of contexts in AI was suggested by John McCarthy in his Turing Award lecture [McCarthy, 1987], as a means to overcome the apparent lack of generality in AI systems. The notion was developed further in a series of papers, see [McCarthy and Buvač, 1994; Buvač *et al.*, 1995; Buvač, 1996] for some recent developments.

The notation $\mathbf{ist}(c, \psi)$, with the reading that ψ is true in the context c, was introduced in [Guha, 1991], where the application is to localized contexts in the CYC knowledge base. We augment this notation by allowing formulas to appear in the first argument of $\mathbf{ist}(\cdots,\cdots)$, and give formal semantics corresponding to the informal reading of $\mathbf{ist}(\phi, \psi)$ as 'ψ is true in the context **described by** ϕ'.

What is a context? The term is used in a variety of senses, and it is doubtful whether any single conception unifies them all. The range of phenomena found under the heading of 'context', includes:

- Mathematical theories, e.g. vector spaces. Mathematicians are in the habit of developing highly abstract, self-contained context descriptions, with precise conditions on when and how they are applicable.
- Conversational settings, that determine the interpretation of indexicals like 'we', 'here', and 'this'. Context accumulates with discourse.
- University politics during the annual budget discussions. Standards and patterns of interaction between colleagues tend to be influenced by this context.

In this paper we present logical tools for reasoning about contexts. We define a formula language with names for contexts, give a fibred model structure and semantics, devise a sound and complete axiomatic system for it, and investigate properties of the resulting logic. Furthermore, we present a general method for deriving logics of context by self fibring.

2 A LOGIC OF IMPLICIT AND EXPLICIT CONTEXT

We define a language of formulas inductively, starting from a set A of atomic formulas, and a nonempty set C of context names. We take both of these

P. Bonzon, M. Cavalcanti and R. Nossum (eds.), Formal Aspects of Context, 193–209.

to be countably infinite in number, although in any given discourse only a finite number of each will actually be used.

DEFINITION 1 (The language L). The set L of formulas is defined as:

$$L ::= A|\neg L|L \rightarrow L|\mathbf{ist}(C, L)|\mathbf{ist}(L, L)$$

The symbols $\neg$ and $\rightarrow$ stand for negation and material implication, respectively. The other classical connectives $\wedge, \vee, \leftrightarrow$ etc., and the constant atoms $\top$ and $\bot$ can be added to the language in the usual way.

For example, C might contain contexts like

ctp 'the College end-of-term party'

abr 'the annual budget round'

wdd 'the wedding of the Dean's daughter'

and A might contain the propositions

sn 'it is snowing outside'

Dh 'the Dean of the College is happy'

rah 'road accident rate is high'

Then the following are examples of well-formed formulas, and our logic will allow all of these to be simultaneously satisfiable. We annotate them by the story of some fictitious College and its Dean, whose mood is known to vary with context. It is wintertime:

$$sn$$

$$\mathbf{ist}(sn, rah)$$

The annual budget negotiations take place at this time of year, and the Dean is usually in a sombre and pessimistic mood:

$$\mathbf{ist}(abr, \neg Dh)$$

During a break in the negotiations, invitations to the College end-of-term party are handed out, and the Dean's mood lifts temporarily:

$$\mathbf{ist}(abr, \mathbf{ist}(ctp, Dh))$$

At the party, there is general merriment:

$$\mathbf{ist}(ctp, Dh)$$

Between the tango and the rumba, a young lecturer draws the Dean aside to argue the financial needs of a particular research project. The Dean's reply is non-committal, with admonitions of austerity in these troubled times for the College:

$$\mathbf{ist}(ctp, \mathbf{ist}(abr, \neg Dh))$$

A few days later the Dean's daughter gets married. When somebody makes a jocular comparison between the lavishness of the wedding reception and the frugality of College budgets, the Dean dismisses this with a laugh and proposes a toast to happiness and prosperity:

$$\mathbf{ist}(wdd, \mathbf{ist}(abr, Dh))$$

The last two examples illustrate nested contexts, and show that the truth of a formula depends on the sequence of surrounding contexts.

2.1 Fibred models of context formulas

To make precise the meaning of formulas, we give a model framework within which to interpret elements of the language L. Our models have a 'possible-worlds' structure, that is to say the truth value of a formula is evaluated at each of a set of points in each model.

Each point, or possible world, represents a coherent interpretation of the propositional language fragment, and the points are connected in a way that facilitates interpretation of $\mathbf{ist}(\ldots,\ldots)$ formulas.

The interpretation of a formula $\mathbf{ist}(c, \phi)$ at a point w depends on the interpretation of ϕ at another point, which is connected to w by a fibre labelled with the context c.

For each point, there is a bundle of fibres connecting it with other pints, one fibre per context. Formulas of the form $\mathbf{ist}(\phi, \psi)$ are interpreted at a point by looking at all other points connected to it by a fibre in the bundle.

Formally, a model in this framework is a triple $M = \langle W, F, V\rangle$, where

- $W \neq \emptyset$,
- $F : W \times C \to W$, and
- $V : W \times A \to 2$.

The F component imposes a graph structure on W, where the edges are labelled by contexts. Paths in the graph correspond to sequences of nested contexts.

If $\vec{\sigma}$ is a finite sequence $\langle c_1, \ldots, c_k\rangle$ of contexts, we shall write $\mathbf{ist}(\vec{\sigma}, \phi)$ as an abbreviation for $\mathbf{ist}(c_1, \ldots \mathbf{ist}(c_k, \phi)\ldots)$, and $F(w, \vec{\sigma})$ as an abbreviation for $F \ldots F(w, c_1)\ldots, c_k)$. When $\vec{\sigma}$ is empty, $\mathbf{ist}(\vec{\sigma}, \phi)$ just ϕ, and $F(w, \vec{\sigma})$ is just w.

DEFINITION 2 ($\models$). The truth value of a formula ϕ at a point $w \in W$ in a model M is denoted $M, w \models \phi$, and defined by the following clauses:

$M, w \models a$ iff $V(w, a)$

$M, w \models \neg\psi$ iff $M, w \not\models \psi$

$M, w \models \phi \to \psi$ iff $M, w \models \phi$ implies $M, w \models \psi$

$M, w \models \mathbf{ist}(c, \psi)$ iff $M, F(w, c) \models \psi$

$M, w \models \mathbf{ist}(\phi, \psi)$ iff $M, w \models \mathbf{ist}(c, \phi \to \psi)$
for all $c \in C$

DEFINITION 3 (Satisfaction). Truth of a formula ϕ at all points of a model M is denoted $M \models \phi$, and we then say that M *satisfies* ϕ.

DEFINITION 4 (Validity). Truth of a formula ϕ at all points in all models, is denoted $\models \phi$, and we then say that ϕ is *valid*.

Every atomic formula $a \in A$ is interpreted as a proposition which is either true or false, and the connectives $\neg$ and $\to$ are interpreted classically. Therefore, all substitution instances of propositional tautologies are valid. A contextualised formula $\mathbf{ist}(c, \psi)$ is interpreted as truth of ψ in context c, and $\mathbf{ist}(\phi, \psi)$ is interpreted as truth of ψ in the context described by ϕ.

Strictly speaking, the symbol **ist** is standing in for a pair of distinct symbols corresponding to its two different interpretations, but it will always be clear from syntax which one is intended.

2.2 Deductive system

We shall define a deductive system which will turn out to determine exactly the set of valid formulas in the model framework we just presented. This is useful, since it gives us two alternative perspectives on the set of valid formulas, a semantic view and a syntactic one.

The axiomatic system consists of axiom schemas and rules of inference.

DEFINITION 5 ($\vdash$). We say that ϕ is a *theorem*, and write $\vdash \phi$, if ϕ is an instance of an axiom schema, or follows from other theorems by application of a rule of inference.

DEFINITION 6 (Soundness). An axiom is said to be *sound* with respect to a model framework if it is valid, and a rule of inference is said to be sound if it takes valid premises to valid conclusions.

If all axioms and rules are sound, we say the whole system is sound, and we then have

$$\vdash \phi \text{ implies } \models \phi$$

for all formulas ϕ.

DEFINITION 7 (Completeness). If every valid formula is a theorem;

$$\models \phi \text{ implies } \vdash \phi$$

then the system is said to be *complete.*

Completeness is usually trickier to establish than soundness, and in the proof of completeness below we shall need some terminology concerning theoremhood of formulas, their negations, and sets of formulas:

DEFINITION 8 (Consistency). We say a formula ϕ is *consistent* iff $\nvdash \neg\phi$, thus *inconsistent* iff $\vdash \neg\phi$. A finite set of formulas is said to be consistent iff the conjunction of its members is consistent, and an infinite set is consistent iff all its finite subsets are consistent. A formula ϕ is *consistent with* a set Γ according to the consistency of $\Gamma \cup \{\phi\}$.

Clearly, when Γ is a consistent set, ϕ is consistent with Γ iff $\neg\phi$ is inconsistent with Γ.

Note that by comparison, we did *not* extend the notion of truth, satisfaction and validity to sets of formulas, in particular not to infinite sets. Thus our notion of completeness can be rephrased as 'every consistent formula is true at some point in some model', but this does not carry over to infinite sets of formulas. There exists a stronger notion of completeness, which does not hold in this logic, the technical reason being that it is not compact. Our only use of infinite sets of formulas is in maximal consistent sets, to be defined next. These will in turn be used to construct a particular model during the proof of completeness.

DEFINITION 9 (Maximality). A consistent set Γ is *maximal* iff, for all formulas ϕ, consistency of $\Gamma \cup \{\phi\}$ implies $\phi \in \Gamma$.

In the proof below, we use the following standard properties of maximal consistent sets of formulas. For a full treatment of maximal consistent sets and their properties, consult e.g. [Chellas, 1980].

- Γ is maximal consistent if it satisfies these three conditions:
  - Contains all theorems: $\vdash \phi$ implies $\phi \in \Gamma$
  - Separates formulas from their negations: $\neg\phi \in \Gamma$ iff $\phi \notin \Gamma$
  - Is propositionally closed: $\phi \in \Gamma$ and $\phi \to \psi \in \Gamma$ implies $\psi \in \Gamma$
- if Γ is maximal consistent, then
  - $\neg\phi \in \Gamma$ iff $\phi \notin \Gamma$
  - $\phi \to \psi \in \Gamma$ iff $\phi \in \Gamma$ implies $\psi \in \Gamma$
  - If $\vdash \phi \to \psi$ then $\phi \in \Gamma$ implies $\psi \in \Gamma$

Axiom schemas

$\mathcal{PT}$ All instances of propositional tautologies.

$\mathcal{DD}$ $\mathbf{ist}(c, \neg\phi) \leftrightarrow \neg\mathbf{ist}(c, \phi)$

$\mathcal{K}$ $\mathbf{ist}(c, \phi \rightarrow \psi) \rightarrow (\mathbf{ist}(c, \phi) \rightarrow \mathbf{ist}(c, \psi))$

$\mathcal{SP}$ $\mathbf{ist}(\phi, \psi) \rightarrow \mathbf{ist}(c, \phi \rightarrow \psi)$

Inference rules

$\mathcal{MP}$ $\dfrac{\vdash \phi, \vdash \phi \rightarrow \psi}{\vdash \psi}$

$\mathcal{RN}$ $\dfrac{\vdash \phi}{\vdash \mathbf{ist}(c, \phi)}$

$\mathcal{RG}$ $\dfrac{\vdash \phi \rightarrow \mathbf{ist}(\vec{\sigma}, \mathbf{ist}(c, \psi \rightarrow \chi))}{\vdash \phi \rightarrow \mathbf{ist}(\vec{\sigma}, \mathbf{ist}(\psi, \chi))}$ where c does not occur in ϕ.

We now proceed to prove soundness and completeness of this axiomatic presentation of the logic.

Soundness proof

We verify the soundness of $\mathcal{SP}$ and $\mathcal{RG}$ as examples. The other axiom schemas and rules are less complicated, and use the same patterns of reasoning.

$\mathcal{SP}$ We fix some arbitrary model $M = \langle W, F, V \rangle$, and show that $\mathcal{SP}$ is true at all $w \in W$: assuming the antecedent of the implication true: $M, w \models \mathbf{ist}(\phi, \psi)$, we must prove its consequent true for the same M and w: $M, w \models \mathbf{ist}(c, \phi \rightarrow \psi)$. But from the assumption we get: $M, w \models \mathbf{ist}(c, \phi \rightarrow \psi)$ for every $c \in C$, which is sufficient.

$\mathcal{RG}$ Assuming that the premise of the rule is valid in all models: $M, w \models \phi \rightarrow \mathbf{ist}(\vec{\sigma}, \mathbf{ist}(c, \psi \rightarrow \chi))$ for every model $M = \langle W_M, F_M, V_M \rangle$ and every $w \in W_M$, we must prove that the conclusion of the rule is valid in any arbitrary model $N = \langle W_N, F_N, V_N \rangle$, in other words $N, u \models \phi \rightarrow \mathbf{ist}(\vec{\sigma}, \mathbf{ist}(\psi, \chi))$ for every $u \in W_N$. So let us fix some such N and a $u \in W_N$, and assume $N, u \models \phi$, to prove $N, u \models \mathbf{ist}(\vec{\sigma}, \mathbf{ist}(\psi, \chi))$. Now, following the semantical definition, we take an arbitrary context $d \in C$, and prove $N, F_N(t, d) \models \psi \rightarrow \chi$ with $t = F_N(u, \vec{\sigma})$. To this end, we construct a special model M from N as follows: its domain is C^*, the set of sequences of contexts, including the empty sequence

ϵ. Intuitively, such a sequence points out a world in W_N, reachable from u by repeated application of F_N. Distinct paths from u reaching the same world in W_N count as distinct worlds in W_M. The crucial difference between N and M is in the interpretation at u, resp. ϵ, of formulas of the form $\mathbf{ist}(\vec{\sigma}, \mathbf{ist}(c, \ldots))$:

$$W_M = C^*$$
$$F_M(w, x) = wx \text{ except } F_M(\vec{\sigma}, c) = \vec{\sigma}\, d$$
$$V_M(w, a) = V_N(F_N(u, w), a)$$

Now we have $M, \epsilon \models \phi$ since $N, u \models \phi$ and c does not occur in ϕ. Therefore by assumption and $\mathcal{MP}$ we have $M, \epsilon \models \mathbf{ist}(\vec{\sigma}, \mathbf{ist}(c, \psi \to \chi))$. Then by the model conditions $M, F_M(\vec{\sigma}, c) \models \psi \to \chi$, and by construction of M it follows that $N, F_N(t, d) \models \psi \to \chi$, as required.

Completeness proof

To establish the completeness of the axiomatic presentation relative to the class of models, we start with a consistent formula $\delta \in L$, and construct a model $M = \langle W, F, V \rangle$ such that $M, w \models \delta$ for a particular $w \in W$.

The structure of the verifying model will be:

- $W = \{w_0\} \cup \{F(w, c) | w \in W, c \in C\}$, where w_0 is a certain set of formulas, containing δ and constructed as described below,
- $F(w, c) = \{\sigma | \mathbf{ist}(c, \sigma) \in w\}$,
- and $V(w, a)$ iff $a \in w$.

The construction of w_0 proceeds in steps as follows. We start with the set $w_0 := \{\delta\}$, and traverse the whole of L, including more formulas as we go: L is clearly enumerable since A and C are, so we fix some enumeration $L = \langle \lambda_1, \lambda_2, \ldots \rangle$. Now we consider each λ_i in turn, and if λ_i is consistent with w_0, then we add λ_i to w_0. If furthermore λ_i is of the form $\mathbf{ist}(\vec{\sigma}, \neg\mathbf{ist}(\psi, \chi))$, then we also add $\mathbf{ist}(\vec{\sigma}, \neg\mathbf{ist}(c_*, \psi \to \chi))$ to w_0, where c_* is chosen as a member of C that does not occur in any member of w_0. Since at each stage of the process w_0 is finite, while C is countably infinite, this is always feasible.

In order to establish completeness we need some lemmas:

LEMMA 10. *w_0 is a maximal consistent set.*

Proof. By induction on the number of steps in the construction process. w_0 is consistent to begin with, and we show that each addition to it preserves consistency. Then, in the limit, every finite subset of w_0 will be consistent, therefore w_0 itself will be consistent too. Also it will be maximal, for suppose that $w_0 \cup \{\lambda_i\}$ is consistent for some $\lambda_i \in L$, then $\lambda_i \in w_0$, since it was added in step i of the process.

Addition of λ_i in the i'th step is only done if it preserves consistency, so it remains to show that, after adding $\mathbf{ist}(\vec{\sigma}, \neg\mathbf{ist}(\psi, \chi))$ consistently, adding $\mathbf{ist}(\vec{\sigma}, \neg\mathbf{ist}(c_*, \psi \to \chi))$ to w_0 also preserves consistency.

To see this, suppose for contradiction that $\mathbf{ist}(\vec{\sigma}, \neg\mathbf{ist}(c_*, \psi \to \chi))$ is inconsistent with w_0, in other words,

$$\vdash \neg(\phi \wedge \mathbf{ist}(\vec{\sigma}, \neg\mathbf{ist}(c_*, \psi \to \chi)))$$

where ϕ is the (finite) conjunction of members of w_0 after adding λ_i consistently. Equivalently

$$\vdash \phi \to \neg\mathbf{ist}(\vec{\sigma}, \neg\mathbf{ist}(c_*, \psi \to \chi))$$

or equivalently, by repeated application of $\mathcal{DD}$,

$$\vdash \phi \to \mathbf{ist}(\vec{\sigma}, \mathbf{ist}(c_*, \psi \to \chi))$$

But then by $\mathcal{RG}$:

$$\vdash \phi \to \mathbf{ist}(\vec{\sigma}, \mathbf{ist}(\psi, \chi))$$

since c_* is chosen so as to not occur in ϕ. By repeatedly applying $\mathcal{DD}$ again, we get

$$\vdash \phi \to \neg\mathbf{ist}(\vec{\sigma}, \neg\mathbf{ist}(\psi, \chi))$$

or equivalently

$$\vdash \neg(\phi \wedge \mathbf{ist}(\vec{\sigma}, \neg\mathbf{ist}(\psi, \chi)))$$

But this contradicts the consistency of λ_i with w_0, so it follows that consistency of w_0 is preserved at every step. This proves that w_0 as constructed above is a maximal consistent set. ■

Next we establish that

LEMMA 11. *$F(w, c)$ is a maximal consistent set whenever w is.*

Proof. By the properties of maximal consistent sets, this follows from these three items:

- $F(w, c)$ contains all theorems: Suppose $\vdash \phi$. Then $\vdash \mathbf{ist}(c, \phi)$ by $\mathcal{RN}$, so $\mathbf{ist}(c, \phi) \in w$ since w is a maximal consistent set. Then it follows that $\phi \in F(w, c)$ by the definition of F.

- $F(w, c)$ separates formulas from their negations, i.e. $\phi \notin F(w, c)$ iff $\neg\phi \in F(w, c)$:

  Expanding the definition of the former we obtain: $\mathbf{ist}(c, \phi) \notin w$ which is equivalent to $\neg\mathbf{ist}(c, \phi) \in w$ by the fact that w is maximal and consistent. By $\mathcal{DD}$ this is equivalent to $\mathbf{ist}(c, \neg\phi) \in w$, which again by definition of F is equivalent to $\neg\phi \in F(w, c)$.

- $F(w,c)$ is propositionally closed, i.e. if $\phi \in F(w,c)$ and $\phi \rightarrow \psi \in F(w,c)$ then $\psi \in F(w,c)$:

  Suppose $\phi \in F(w,c)$, i.e. by definition $\mathbf{ist}(c,\phi) \in w$, and suppose also $\phi \rightarrow \psi \in F(w,c)$, which develops into $\mathbf{ist}(c,\phi \rightarrow \psi) \in w$. We must show $\psi \in F(w,c)$, which means $\mathbf{ist}(c,\psi) \in w$. But this follows from $\mathcal{K}$ and the fact that w is a maximal consistent set.

This proves that the set $F(w,c)$ is maximal and consistent whenever w is, and by induction the previous two lemmas prove that all $w \in W$ are maximal consistent sets. ■

LEMMA 12. $M,w \models \phi$ *iff* $\phi \in w$

Proof. This is proved by induction on the structure of the formulas.

- The atomic case follows directly from the definition of V

- $\neg\phi$: $M,w \models \neg\phi$ iff (by definition) $M,w \not\models \phi$ iff (by induction) $\phi \notin w$ iff (since w is maximal consistent) $\neg\phi \in w$

- $\phi \rightarrow \psi$: $M,w \models \phi \rightarrow \psi$ iff (by definition) $M,w \models \phi$ implies $M,w \models \psi$ iff (by induction) $\phi \in w$ implies $\psi \in w$ iff (since w is maximal consistent) $\phi \rightarrow \psi \in w$

- $\mathbf{ist}(c,\chi)$: By definition, $M,w \models \mathbf{ist}(c,\chi)$ iff $M,F(w,c) \models \chi$, equivalent by induction to $\chi \in F(w,c)$, which by definition of F is equivalent to: $\mathbf{ist}(c,\chi) \in w$.

- $\mathbf{ist}(\phi,\psi)$: By definition, $M,w \models \mathbf{ist}(\phi,\psi)$ iff for every $c \in C$, $M,w \models \mathbf{ist}(c,\phi \rightarrow \psi)$. By induction, this is equivalent to: for every $c \in C$, $\mathbf{ist}(c,\phi \rightarrow \psi) \in w$.

  Now suppose that $\mathbf{ist}(\phi,\psi) \in w$. Then, by $\mathcal{SP}$ and the fact that w is a maximal consistent set, $\mathbf{ist}(c,\phi \rightarrow \psi) \in w$ for any c, which as seen above is equivalent to $M,w \models \mathbf{ist}(\phi,\psi)$.

  Suppose on the contrary that $\mathbf{ist}(\phi,\psi) \notin w$. Then, since w is a maximal consistent set, $\neg\mathbf{ist}(\phi,\psi) \in w$. By construction of the model,

  - $w = F(w_0,\vec{\sigma})$ for some (possibly empty) sequence $\vec{\sigma}$ of contexts
  - $\mathbf{ist}(\vec{\sigma},\neg\mathbf{ist}(\phi,\psi)) \in w_0$
  - $\mathbf{ist}(\vec{\sigma},\neg\mathbf{ist}(c_*,\phi \rightarrow \psi)) \in w_0$ for some select c_*

  It follows that $\neg\mathbf{ist}(c_*,\phi \rightarrow \psi) \in w$, which as seen above is equivalent to $M,w \not\models \mathbf{ist}(\phi,\psi)$.

This completes the proof of the lemma. ■

THEOREM 13. $M, w_0 \models \delta$

Proof. Follows from the previous lemma, since $\delta \in w_0$. ■

We have shown that every consistent formula is true somewhere, so we have completeness.

2.3 Further properties of the logic

Now that we have established a sound and complete axiomatic basis for the logic, we comment on a few additional properties of it, in the form of sound axiom schemas and rules of inference.

The soundness of each schema and rule is sufficient to ensure their dependence on the complete axiomatic basis. Soundness is straightforward to verify every case, and we omit the proofs.

Axiom schemas

$\mathcal{TC}$ $\quad \mathbf{ist}(\top, \phi) \to \mathbf{ist}(c, \phi)$

$\mathcal{DD'C}$ $\quad \neg(\mathbf{ist}(\top, \neg\phi)) \to (\mathbf{ist}(\phi, \neg\psi) \leftrightarrow \neg\mathbf{ist}(\phi, \psi))$

The least specific description of a context is $\top$, which describes everything by virtue of being true no matter what. $\mathcal{TC}$ expresses this, and $\mathcal{DD'C}$ says that whenever ϕ is not false in every context, then whatever is false in the contexts described by ϕ, is not true there. The restriction on ϕ is to avoid empty quantification.

$\mathcal{REF}$ $\quad \mathbf{ist}(\phi, \phi)$

$\mathcal{TRA}'$ $\quad \mathbf{ist}(x, \phi) \to (\mathbf{ist}(\phi, \psi) \to \mathbf{ist}(x, \psi))$, for $x \in C$ or $x \in L$.

This shows that **ist** is a partial preorder on L.

$\mathcal{I}'A$ $\quad (\phi \to \psi) \to (\mathbf{ist}(\psi, \chi) \to \mathbf{ist}(\phi, \chi))$

$\mathcal{IC}$ $\quad (\phi \to \psi) \to (\mathbf{ist}(x, \phi) \to \mathbf{ist}(x, \psi))$ for $x \in C$ or $x \in L$.

These axioms express that **ist** is antitone in its first coordinate and monotone in its second one.

Rules of inference

$$\mathcal{REA} \quad \frac{\vdash \phi \leftrightarrow \psi}{\vdash \mathbf{ist}(\phi, \chi) \leftrightarrow \mathbf{ist}(\psi, \chi)}$$

$$\mathcal{RKC} \quad \frac{\vdash \wedge\phi_i \to \psi}{\vdash \wedge\mathbf{ist}(x, \phi_i) \to \mathbf{ist}(x, \psi)} \text{ for } x \in C \text{ or } x \in L.$$

These two rules show that, in the terminology of [Chellas, 1980], if we look at **ist** as a binary modality, it is *classical* in its first coordinate and *normal* in its second one. It shares these properties with the class of conditional logics investigated there, and for which the model framework was minimal models. We feel that the class of models we have devised here is simpler and more intuitive.

3 CONTEXT SYSTEMS THROUGH SELF FIBRING OF PREDICATE LOGICS

We now proceed to give a more general treatment of the $\mathbf{ist}(\cdots,\cdots)$ predicate. This section will show how various formal systems of context can be naturally presented as self fibring of predicate logics. The term self fibring refers to fibres that connect models of the same logic.

There is a strong similarity of context formalism with that of Labelled Deductive Systems (*LDS*) [Gabbay, 1996b][1]. In fact, the machinery of *LDS* can provide the formal logical framework for theories of context.

An algebraic LDS is a triple $\Lambda = (\Delta, L, R)$ where Δ is a set of formulas and L is an algebra of labels. These give rise to so-called *declarative units*, which are pairs $\lambda : F$ consisting of a label and a formula. R is a deductive discipline for deriving declarative units.

In [McCarthy and Buvač, 1994] and elsewhere, asserting $\mathbf{ist}(\phi, \psi)$ in context c is denoted by

$$c : \mathbf{ist}(\phi, \psi)$$

which is readily identified with a declarative unit in an LDS with labels as a part of the formula language.

The notion of fibred semantics was introduced in 1991 to give semantics for *LDS* and from it the notions of fibring logics and self fibring emerged, grew and developed into [Gabbay, 1996a]. It gives a fuller account of fibred semantics than is possible here.

3.1 The system B *of context*

We begin by defining a certain self fibred predicate logic which will naturally specialise to the system of [Buvač, 1996].

Let $\mathbf{L}_1$ be a copy of predicate logic and let $\mathbf{ist}(x, y)$ be a binary predicate of $\mathbf{L}_1$. Let $\mathbf{L}_0$ be the language of the predicate logic with the parameterised unary predicate $\lambda y \mathbf{ist}(x, y)$.

Consider the self fibred language $\mathbf{L}_0[\mathbf{L}_1]$. This means we can repeatedly substitute fibred formulas for the y coordinate of $\mathbf{ist}(x, y)$.

---

[1] Gabbay's *LDS* position paper was presented in Logic Colloquium 90 [Gabbay, 1993]. A first draft of the *LDS* book was available in 1989 [Gabbay, 1989].

Consider a fibred model for this language. As is shown in [Gabbay, 1996a], we can take as models for this language structures of the form $\mathbf{n} = (S, D, a, \mathbf{F}, h, g)$, satisfying certain conditions.

In such models,

- S is the set of labels naming classical models.
- D is the domain which we assume for the purpose of this section to be constant (the same) for all labels (classical models)
- $a \in S$ is the actual label
- $\mathbf{F}$ is the fibring function associating with each $X \subseteq D$ and $t \in S$ a set of labels $\mathbf{F}(X,t) \subseteq S$
- h is the interpretation associating for each $t \in S$ and each **m**-place predicate P a subset $h(t,P) \subseteq D^m$
- g is a rigid assigment giving for each variable or constant x of the language an element $g(x) \in D$.

Satisfaction is defined as in [Gabbay, 1996a] with the clause for $\mathbf{ist}(x,\phi)$ being:

- $t \vdash \mathbf{ist}(x,\phi)$ iff for all $s \in \mathbf{F}(X_{t,x},t)$, we have $s \vdash \phi$, where $X_{t,x} = \{y \mid t \vdash \mathbf{ist}(x,y)\}$.

Let $\mathcal{M}^t_x = \mathbf{F}(X_{t,x},t)$. We have

- $t \vdash \mathbf{ist}(Xx,\phi)$ iff $s \vdash \phi$, for all $s \in \mathcal{M}^t_x$.

REMARK 14. **The system $\mathbf{B}_0$.** Let $\mathbf{B}_0$ be the system semantically defined above. $\mathbf{B}_0$ can be axiomatised as shown in [Gabbay, 1996a]. This gives us a straightforward basic system of context. The system of [Buvač, 1996], which we shall call $\mathbf{B}_u$, is obtained by further simplifying assumptions on the semantics. This we discuss below.

We now examine what kind of simplifying assumptions we can have on $\mathbf{F}$ and on $h(t, \mathbf{ist})$.

OPTION 1

We can assume that $\mathcal{M}^t_x$ is independent of t. This can be achieved by assuming that $X_{t,x}$ does not depend on t and further that $\mathbf{F}$ does not depend on t. The first assumption means that **ist** is a rigid predicate, i.e. there exists a relation $R_{\mathbf{ist}} \subseteq D^2$ such that for all t $h(t,\mathbf{ist}) = R_{\mathbf{ist}}$. The second assumption means that $\mathbf{F}(X,t)$ is rigid, i.e. does not depend on t. To appreciate the implications of this option, consider now the evaluationof $t \vdash \mathbf{ist}(x_1, \mathbf{ist}(x_2,\phi))$. Expanding the definition we get iff $(\forall s \in \mathcal{M}^t_{x_1})(\forall r \in \mathcal{M}^s_{x_2})(r \vdash \phi)$, which we can write compactly as $(\forall r \in \mathcal{M}^t_{(x_1,x_2)})(r \vdash \phi)$, where $r \in \mathcal{M}^t_{(x_1,x_2)}$ iff (definition) there exists an s such that $(s \in \mathcal{M}^t_{x_1} \wedge r \in \mathcal{M}^s_{x_2})$.

Since we assumed that $\mathcal{M}^t_x$ does not depend on t at all, we get

- $t \vdash \mathbf{ist}(x_1, \mathbf{ist}(x_2, \phi))$ iff for all $r \in \mathcal{M}_{(x_1,x_2)}, r \vdash \phi$.

Let us take a closer look at $\mathcal{M}_{(x_1,x_2)}$ in case $\mathcal{M}$ does not depend on h. $r \in \mathcal{M}_{(x_1,x_2)}$ iff $\exists s(s \in \mathcal{M}_{x_1} \wedge r \in \mathcal{M}_{x_2})$. Assuming that $\mathcal{M}_{x_1}$ is not empty we get $\mathcal{M}_{(x_1,x_2)} = \mathcal{M}_{x_2}$.

This immediately entails that $\mathbf{ist}(x_1, \mathbf{ist}(x_2, \phi)) = \mathbf{ist}(x_2, \phi)$, which is the flatness property found in e.g. [Buvač, 1996].

OPTION 2

We note the expression $t \vdash \mathbf{ist}(x, \phi)$. x ranges over D and t ranges over S. We can opt to make **ist** a two sorted predicate and let x range over S. This assumption does not imply any loss of generality since there is no other 'serious' substitution in the x-coordinate of **ist** and the second coordinate of **ist** is not affected. Choosing this option brings us closer to the system of [Buvač, 1996].

We can simplify further by letting $S = D$, and we need not have a two sorted system. This can also be done without loss of generality. The two choices give the same system in essence.

If we adopt the above two options, and further assume no function symbols, what we get is the self fibring of predicate logic into a family of unary predicates $\mathbf{ist}_t(y)$ indexed by the labels $t \in S$, whose models have the form $(S, D, a, \mathbf{F}, h, g)$ where both $\mathbf{F}$ and $h(t, \mathbf{ist}_s)$ are *rigid*.

Using results from [Gabbay, 1996a], we immediately get axiomatisability and decidability, summarised in the definition and theorem below.

DEFINITION 15. The context system $\mathbf{B}_u$. Let $\mathbf{L}_1$ be the monadic language with a family of unary predicates $\mathbf{ist}_t$, $t \in C$ and $\mathbf{L}_2$ be the classical predicate language. Let $\mathbf{L}_1[\mathbf{L}_2]$ be the self fibred language obtained by allowing repeated substitution from $\mathbf{L}_2$ into $\mathbf{L}_1$, as defined in [Gabbay, 1996a]. Let $\mathbf{B}$ be the Hilbert system for this language with the following axioms and rules.

1. All axioms and rules of free predicate logic with non-denoting constants $\mathcal{T}$. These include

   - All substitution instances of truth functional tautologies
   - $\dfrac{\vdash \phi, \vdash \phi \to \psi}{\vdash \psi}$
   - $\forall x(\phi(x) \wedge \psi(x)) \to \forall x\phi(x) \wedge \forall x\psi(x)$
   - $\forall x\phi(x) \wedge \forall x(\phi(x) \to \psi(x)) \to \forall x\psi(x)$
   - $\dfrac{\vdash \phi \to \psi(x)}{\vdash \phi \to (x)\psi(x)}$ where x is not free in ϕ

- $\forall x \forall y \phi \to \forall y \forall x \phi$
- $\forall u(\psi \wedge \forall x \phi(x) \to \phi(u))$.

2. Modal rules $\mathcal{K}$

- $\dfrac{\vdash \wedge \phi_i \to \phi}{\vdash \mathbf{ist}(t, \wedge \phi_i) \to \mathbf{ist}(t, \phi)}$
- $\forall x\, \mathbf{ist}(t, \phi(x)) \to \mathbf{ist}(t, \forall x\, \phi(x))$.

3. Axiom for the rigidity of $\mathbf{F}$.

- $\forall x[\mathbf{ist}(t_1, x) \leftrightarrow \mathbf{ist}(t, \mathbf{ist}(t_2, x))] \to [\mathbf{ist}(t_1, \phi) \leftrightarrow \mathbf{ist}(t, \mathbf{ist}(t_2, \phi))]$.

4. Axiom for the rigidity of **ist**

- $\mathbf{ist}(t, x) \to \mathbf{ist}(s, \mathbf{ist}(t, x))$
- $\neg\mathbf{ist}(t, x) \to \mathbf{ist}(s, \neg\mathbf{ist}(t, x))$

REMARK 16. Here we are applying the ready-made machinery of [Gabbay, 1996a] to the particular case of the context langauge.

THEOREM 17 (Completeness theorem for $\mathbf{B}_u$). *$\mathbf{B}_u$ is complete for the class of fibred models with* **F** *and* **ist** *rigid.*

Proof. see [Gabbay, 1996a]. ■

THEOREM 18 (Decidability theorem for $\mathbf{B}_u$). *The tight fragment of monadic $\mathbf{B}_u$ is decidable.*

Proof. See [Gabbay, 1996a]. ■

Note that by monadic $\mathbf{B}_u$, we mean we have only monadic predicates in the classical language and by *tight* fragment we mean that we substitute in $\mathbf{ist}(x, y)$ in the y position only wffs with at most y free. Otherwise we get undecidability!

We compare with [Buvač, 1996]. Our models for $\mathbf{B}_u$ are practically identical but for equality, but that can be added without any problem. Other minor differences: we use letters t for models instead of $\mathcal{A}$, we write $t \vdash \phi$ for evaluation, as compared with $t \vdash s : \phi$, where the s is redundant as can be seen from [Buvač, 1996] by inspection.

The following axioms are given in [Buvač, 1996]:

1. Classical predicate calculus axioms and rules

2. Propositional properties of contexts:

$\mathcal{K}$ $\;k : \mathbf{ist}(k', \phi \to \psi) \to (\mathbf{ist}(k', \phi) \to \mathbf{ist}(k', \psi))$

(Δ) $k : \mathbf{ist}(k_1, \mathbf{ist}(k_2, \phi)) \vee \mathbf{ist}(k_1, \neg\mathbf{ist}(k_2, \phi))$

(Flat) $k : \mathbf{ist}(k_1, \mathbf{ist}(k_2, \phi)) \leftrightarrow \mathbf{ist}(k_2, \phi)$

$$\text{(Enter)}\ \frac{s : \mathbf{ist}(t, \phi)}{t : \phi}$$

$$\text{(Exit)}\ \frac{t : \phi}{s : \mathbf{ist}(t, \phi)}$$

The last two rules correspond to saying

$t : \phi$ is the same as $\mathbf{ist}(t, \phi)$

$t : (s : \phi)$ equals $s : \phi$

3. Quantificational properties of contexts:

(BF) $k : \forall x(\mathbf{ist}(k', \phi) \rightarrow \mathbf{ist}(k', \forall x \phi)$

$(\mathbf{ist} =)$ $k : t_1 = t_2 \leftrightarrow \mathbf{ist}(k', t_1 = t_2)$

Let us compare this axiom system with our own system defined above: Group axioms (1) of our system yields all the classical axioms. The $\mathcal{K}$ group of rules yield context axioms $\mathcal{K}$, (Δ) and (BF). The axiom (Flat) is obtained from our rigidity axioms. The rules (Enter) and (Exit) are obviously conservative additional axioms, and we can add them if desired. The reader should bear in mind that we did not invent our axioms for $\mathbf{B}_u$ as a specific system. We have a general self fibring mechanism that is axiomatised and it specialises to certain axioms for the case of $\mathbf{B}_u$. It will specialise to other axioms should the reader choose a different context logic.

Let us list what we believe self fibring methodology can offer to context theories.

1. We provide a wide framework for general self fibring. This is useful since in AI logics and applications there is a lot of use of substitutions of formulas inside formulas.

2. More specifically for the McCarthy–Buvač systems with the $\mathbf{ist}(x, y)$ predicate, we can use the methodology to allow the use of expressions of the form $\mathbf{ist}(\phi, \psi)$, where the context is given by wffs of possibly another logic and language. We can describe contexts in a logic (not only name them) and reason logically about context. We can even allow statements on the context which may involve considerations of what happens in them, i.e. what kinds of wffs ψ hold in them (i.e. $\mathbf{ist}(\phi, \psi)$ holds).

3. Given two known logics, one for L_1 wffs and one for L_2 wffs, we can let the L_1-language serve as context for the L_2-language, by using the $\mathbf{ist}(\phi, \psi)$ predicate. In this way, we have a way to express under what

conditions known properties of the components systems (e.g. decision procedure, proof theory) transfer to the combined context system with $\mathbf{ist}(\phi, \psi)$.

4 COMPARISON WITH THE LITERATURE

In comparison with the propositional $\mathbf{ist}(c, \psi)$ of [Guha, 1991] and the two-sorted quantified version in [Buvač, 1996], our logic of $\mathbf{ist}(\phi, \psi)$ introduces new notation and develops a new logic for it. Where [Buvač, 1996] admits arbitrary quantification over context variables, our system of section 2 harnesses generalization over contexts in a two-layered multi-modal system, the semantics of one modality quantifying over a set of other modalities. In spite of its richness of expression, we obtained a simple axiomatic characterization of this system.

van Benthem [1996] reviews diverse uses of context, and suggests that the term denotes a convenient methodological fiction, rather than a well-defined ontological category. He finds that admitting full quantification over context variables swamps the language with extraneous elements, and trades them in for greater technical economy. This recalls our position in section 2. His proposal is for an indexing scheme, where each language element can be decorated with an index specifying an intended context for evaluation. This results in a scheme where transition between contexts has a natural expression.

Our system $\mathbf{B}_u$, which we obtained above by self fibring, subsumes the system of [Buvač, 1996], except for minor differences discussed in the previous section.

In [Besnard and Tan, 1995], contextual reasoning is formalized by indexing formulas with sets of formulas, the index set specifying the context. A natural deduction system is given, and a connection with default logic is made via an operator on assumptions. The sets denoting contexts are arbitrary, whereas in our $\mathbf{ist}(\phi, \psi)$, the context is given indirectly by ϕ.

ACKNOWLEDGEMENTS

Thanks to the anonymous referees for their comments on a previous version of this paper, and to Hans-Jürgen Ohlbach for useful and enjoyable discussions.

Dov M. Gabbay
King's College, London.

Rolf Nossum
Agder College, Norway.

REFERENCES

[Benthem, 1996] J. van Benthem. Changing contexts and shifting assertions. In A. Aliseda-Llera, R. van Glabbeek, and D. Westerståhl, editors, *Proceedings 4th CSLI Workshop in Logic, Language and Computation*. CSLI Publications, 1996.

[Besnard and Tan, 1995] P. Besnard and Y.-H. Tan. A modal logic with context-dependent inference for non-monotonic reasoning. Technical Report FS-95-02, AAAI, 1995. Part of the Fall Symposium Series.

[Buvač *et al.*, 1995] S. Buvač, V. Buvač and I. A. Mason. Metamathematics of context. *Fundamenta Informaticae*, 23(3), 1995.

[Buvač, 1996] S. Buvač. Quantificational logic of context. In *Proceedings of the Thirteenth National Conference on Artificial Intelligence*, 1996.

[Chellas, 1980] B. F. Chellas. *Modal Logic: an introduction*. Cambridge University Press, 1980.

[Gabbay, 1989] D. M. Gabbay. Labelled deductive systems, part i. Technical report, Imperial College, 1989.

[Gabbay, 1993] D. M. Gabbay. Labelled deductive systems, a position paper. In *Logic Colloquium 90*, pages 66–89. Lecture notes in logic, vol. 2, Springer Verlag, 1993.

[Gabbay, 1996a] D. M. Gabbay. Fibred semantics. Monograph, Imperial College, 1996.

[Gabbay, 1996b] D. M. Gabbay. *Labelled Deductive Systems, Part I*. Oxford University Press, 1996.

[Guha, 1991] R. V. Guha. *Contexts: A Formalization and Some Applications*. PhD thesis, Stanford University, 1991.

[McCarthy and Buvač, 1994] J. McCarthy and S. Buvač. Formalizing Context (Expanded Notes). Technical Note STAN-CS-TN-94-13, Stanford University, 1994.

[McCarthy, 1987] J. McCarthy. Generality in artificial intelligence. *Comm. of ACM*, 30(12):1030–1035, 1987.

INDEX

APPLIED LOGIC SERIES

1. D. Walton: *Fallacies Arising from Ambiguity.* 1996 ISBN 0-7923-4100-7
2. H. Wansing (ed.): *Proof Theory of Modal Logic.* 1996 ISBN 0-7923-4120-1
3. F. Baader and K.U. Schulz (eds.): *Frontiers of Combining Systems.* First International Workshop, Munich, March 1996. 1996 ISBN 0-7923-4271-2
4. M. Marx and Y. Venema: *Multi-Dimensional Modal Logic.* 1996 ISBN 0-7923-4345-X
5. S. Akama (ed.): *Logic, Language and Computation.* 1997 ISBN 0-7923-4376-X
6. J. Goubault-Larrecq and I. Mackie: *Proof Theory and Automated Deduction.* 1997 ISBN 0-7923-4593-2
7. M. de Rijke (ed.): *Advances in Intensional Logic.* 1997 ISBN 0-7923-4711-0
8. W. Bibel and P.H. Schmitt (eds.): *Automated Deduction - A Basis for Applications.* Volume I. Foundations - Calculi and Methods. 1998 ISBN 0-7923-5129-0
9. W. Bibel and P.H. Schmitt (eds.): *Automated Deduction - A Basis for Applications.* Volume II. Systems and Implementation Techniques. 1998 ISBN 0-7923-5130-4
10. W. Bibel and P.H. Schmitt (eds.): *Automated Deduction - A Basis for Applications.* Volume III. Applications. 1998 ISBN 0-7923-5131-2
    (Set vols. I-III: ISBN 0-7923-5132-0)
11. S.O. Hansson: *A Textbook of Belief Dynamics.* Theory Change and Database Updating. 1999 Hb: ISBN 0-7923-5324-2; Pb: ISBN 0-7923-5327-7
    Solutions to exercises. 1999. Pb: ISBN 0-7923-5328-5
    Set: (Hb): ISBN 0-7923-5326-9
    Set: (Pb): ISBN 0-7923-5329-3
12. R. Pareschi and B. Fronhöfer (eds.): *Dynamic Worlds from the Frame Problem to Knowledge Management.* 1999 ISBN 0-7923-5535-0
13. D.M. Gabbay and H. Wansing (eds.): *What is Negation?* 1999 ISBN 0-7923-5569-5
14. M. Wooldridge and A. Rao (eds.): *Foundations of Rational Agency.* 1999 ISBN 0-7923-5601-2
15. D. Dubois, H. Prade and E.P. Klement (eds.): *Fuzzy Sets, Logics and Reasoning about Knowledge.* 1999 ISBN 0-7923-5911-1
16. H. Barringer, M. Fisher, D. Gabbay and G. Gough (eds.): *Advances in Temporal Logic.* 2000 ISBN 0-7923-6149-0
17. D. Basin, M.D. Agostino, D.M. Gabbay, S. Matthews and L. Viganò (eds.): *Labelled Deduction.* 2000 ISBN 0-7923-6237-3
18. P.A. Flach and A.C. Kakas (eds.): *Abduction and Induction.* Essays on their Relation and Integration. 2000 ISBN 0-7923-6250-0

19. S. Hölldobler (ed.): *Intellectics and Computational Logic.* Papers in Honor of Wolfgang Bibel. 2000 ISBN 0-7923-6261-6
20. P. Bonzon, M. Cavalcanti and Rolf Nossum (eds.): *Formal Aspects of Context.* 2000 ISBN 0-7923-6350-7
21. D.M. Gabbay and N. Olivetti: *Goal-Directed Proof Theory.* 2000 ISBN 0-7923-6473-2

KLUWER ACADEMIC PUBLISHERS – DORDRECHT / BOSTON / LONDON